# The Electric Car Guide: Nissan LEAF

## Your guide to buying and owning a Nissan LEAF

Michael Boxwell

Greenstream Publishing

## Greenstream Publishing

Greenstream Publishing Limited
5 Palmer House
Palmer Lane
Coventry
CV1 1FN
United Kingdom

www.greenstreampublishing.com

Published by Greenstream Publishing 2015
Copyright © Michael Boxwell 2015
First Edition – published April 2015
ISBN 978-1-907670-48-0

# Contents

# Welcome!

This is a book about one of the most important cars of the twenty-first century, the Nissan LEAF.

The LEAF is currently the best-selling electric car in the world. It was one of the very first electric cars to be released by a mainstream manufacturer, and the first affordable family-sized electric car. It was an instant hit with critics and automotive journalists when it was released at the end of 2010. In its first year, it won the biggest automotive awards around the world: the 2011 World Car of the Year, the 2011 European Car of the Year and the 2011-2012 Car of the Year - Japan awards. Critics loved the smooth power delivery, the comfortable and spacious interior and low running costs.

Since then, sales of the LEAF have been increasing as the public perception and confidence in electric cars has grown. Around 150,000 LEAFs have been sold worldwide and sales are growing by around 40% each year. Some countries, including the UK, now have a nationwide charging infrastructure which offers the ability to rapid-charge in around half an hour at motorway service stations and in central city areas, making ownership viable for more and more people.

But what are they like to use on a daily basis? What about the range? How quickly can they be charged up? What are the benefits of electric cars? Can a LEAF be a practical car for every day driving, or do the limitations make them only suitable as an expensive second car?

I've been involved in the electric vehicle industry since 2003 and I've been using electric cars on a daily basis as my own personal transport since 2006. Over the past few years, I've been involved in the design and development of electric cars and battery technologies, worked with a number of major car makers and gained a unique insight into the industry and the technology behind these cars. I drove a number of prototype Nissan electric cars, including the LEAF one year before it went into production and I've owned two Nissan LEAF's of my own.

But I'm also a family man. I use my car for both personal and business use. I drive around 16,000 miles a year and can sometimes end up driving 150 – 200 miles in a single day. For the past few years I've carried out virtually all my driving with a LEAF. It is the only car in our family and carries out its duties as family transport, commuting vehicle and longer distance business driver.

The cost savings have been dramatic. Instead of spending around £250 a month on fuel, I lease my car for just £229 a month and pay around £17 a month extra on electricity. In effect, the money I used to be spending on fuel now buy me a brand new car every three years. The environmental benefits have been significant too. Even taking into account the generation of the electricity and the carbon footprint associated with manufacturing the lithium batteries, I've calculated that I have reduced my carbon footprint by 1750kg (3850 pounds) per year by driving a Nissan LEAF compared to a Prius hybrid or an economical city car.

This book will tell you what it is like to own an electric car in general, and a Nissan LEAF in particular. It won't shy away from the disadvantages and it includes reports from other Nissan LEAF owners as well. Interested in the environmental benefits? The book includes real world side-by-side road tests comparing the Nissan LEAF with a Toyota Prius and a Ford Focus 1.0-litre EcoBoost, measuring the environmental impact of each vehicle.

By the time you have finished reading this book, you'll understand what it is like to own, use and live with one of these exciting cars on a daily basis. You'll understand the benefits, the drawbacks and you'll know whether or not a Nissan LEAF is a suitable car for you.

# An Introduction to Electric Cars

Electric cars are nothing new. In fact, prototype and experimental electric cars were running in the mid-1830s, whilst Englishman Thomas Parker built and sold electric cars to the general public in 1884 and 1885. By the end of the 19th Century, electric cars were on sale from a number of manufacturers across the United States and Europe. Many cities had publicly accessible charging points and electric cars were popular as taxis and doctors cars. Electric cars had a reputation for reliability, quietness and ease of use.

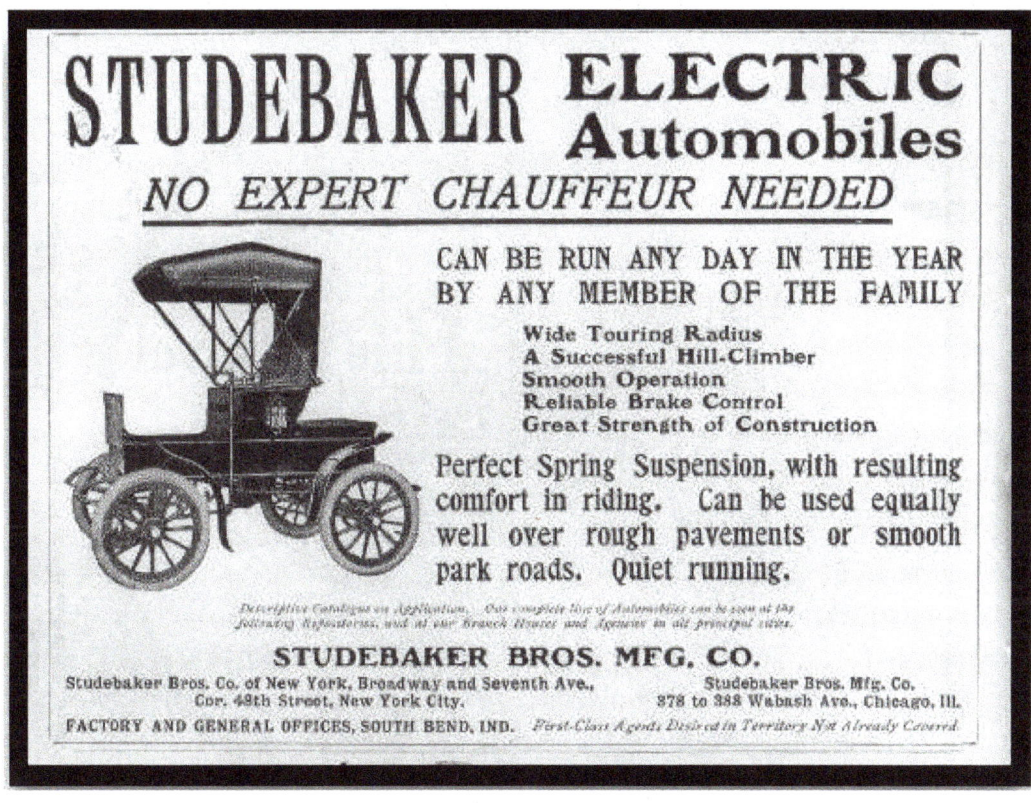

*A 1903 commercial for the Studebaker electric car.*

Yet they were also heavy and expensive to produce. As petroleum became more easily available and internal combustion engines became more reliable and more powerful, the vastly cheaper internal combustion engine car achieved dominance. The electric car faded from the picture and, by the end of the First World War, electric cars had almost completely disappeared.

Electric cars have always had their benefits, but their drawbacks have always been price and performance. And whilst oil was cheap and readily available, as it was for much of the 20th Century, there was never a reason to revisit them. So what changed to make them relevant now? And why should you care?

## Electric cars and the twenty-first century

The 21st Century has seen two significant changes for transportation. The first has been the huge increase in the cost of oil. Back in the late 1990s, a barrel of crude oil could cost as little as $10. By the end of 2000, the price had risen to around $30 a barrel, triggering mass protests across Europe about the rising cost of fuel. Yet prices continued to rise. By the beginning of 2008, the price of crude oil was out of control: shooting up to over $100 a barrel, and increasing until it peaked at over $147 by July.

Fuel was simply becoming unaffordable. With daily price rises and no guarantee that the prices wouldn't rise further, both businesses and the general public were in uproar. Several major transport and courier companies went out of business in the United Kingdom, unable to continue.

Whilst the credit crunch and the resulting worldwide recession was not triggered by oil prices, it is undoubtedly true that the depth of the resulting crash was exacerbated by them. With the drop in trading, demand for oil fell and prices dropped to around $40 a barrel by the end of the year. Since then prices have fluctuated, selling for between $51 and $127 a barrel. Prices can increase or drop by as much as $8 in a single day.

Worldwide demand for oil is now steadily increasing. Emerging nations such as India and China are consuming more and more oil as cars become affordable for the first time. Improvements in fuel economy are offset by people buying bigger and more powerful cars. And whilst oil may be comparatively cheap sometimes, the volatile market fluctuations means that crude oil can double in price in just a few months.

Oil is an incredibly valuable resource. We don't just use it for fuel. Oil is used in nylon and polyester clothing, in plastics, cosmetics, toothpaste, food, macadam (tarmac), cleaning products, soap, medicines and countless other products as well. Fluctuating oil prices does not just affect the cost of transportation, it affects the cost of manufacturing as well.

The second significant change for the 21st Century has been the growing awareness of climate change and the environmental impact of transportation. I'm not going to go into the arguments about climate change here: there are plenty of books that will do that for you. However, it is now generally accepted that the climate change that we are currently experiencing is largely down to man-made pollutants and one of the largest pollutants is transport.

Air quality in our towns and cities is even more directly attributable to road vehicles. The growing awareness of the toxicity of running internal combustion engines – particularly diesel engines – is now having a significant impact on public health. It is estimated that around 29,000 people a year die in the United Kingdom because of traffic-based air pollution. The World Health Organization claims that diesel exhaust fumes are a major cancer risk, as deadly as asbestos, arsenic and mustard gas with similar risks to passive smoking. Young children and the elderly are most at risk. The Mayor of Paris is proposing to ban diesel cars from the central of Paris by 2020.

Of course, electric cars themselves aren't pollution free. The cars themselves don't pollute, but the electricity they are charged from still has to come from somewhere. Even so, the pollution caused by electric cars is still significantly less than conventional cars.

I go into the environmental impact of electric cars in a later chapter, *The Nissan LEAF and the environment*, starting on page 79. Sufficient to say at this stage that electric cars, even if they are powered by fossil fuel power stations, can make a significant contribution to a reduction in pollution, and make our towns and cities nicer places to live and work.

## The advancement of technology

Back in the early 1990s, I was working for a computer manufacturer developing a new mobile computer. In the labs was a prototype Windows-based touch screen smartphone. It was powered by an experimental battery using lithium-ion technology and I was working on the battery management.

Lithium-ion promised much: better packaging, improved energy storage and a longer life. Yet they were also expensive, temperamental, and difficult to control. Get the charging or discharging characteristics wrong and they had a habit of melting, catching fire or exploding. The battery life was around 30 minutes and we had to limit the functionality of the computer so the batteries didn't discharge too fast.

Gradually, the technology improved. Updated chemistry and better battery management meant we could charge and discharge faster and safer. The battery capacity doubled, then doubled again in a matter of a few months. The batteries became lighter, easier to handle and less susceptible to overheating. Battery performance simply got better and better.

Fast forward two decades to 2014, and I am sitting in a meeting room with a team of engineers, scientists, researchers and designers. We've just signed off the phase one development of a new generation of electric vehicle battery pack. A battery pack that is 30% lighter, 30% more compact and 30% cheaper than existing electric vehicle batteries. A thought occurs to me. A quick calculation on a scrap of paper reveals the startling statistic that our new battery was approximately 1/100th of the cost of the original lithium-ion battery I worked with back in 1994, yet had twenty-five times the capacity. In the intervening period, the rate of development had never slowed down.

Likewise, electric motors, motor control systems and other technology used in an electric car has also developed rapidly over the past twenty years. With only three moving parts, electric motors now run at between 90-95% efficiency, compared to 35-42% efficiency of the latest generation of conventional engines. They weigh less than one quarter of a small internal combustion engine, are around one quarter of the size, yet provide much higher levels of torque.

## The changing face of transport

The role of the car in society is changing. It is a subtle shift of perception, but one with significant repercussions on how we will view cars in the future.

Owning and running a car used to be relatively cheap. Wind the clock back twenty years and fuel was cheap, car insurance was cheap and, even if you were a young driver with a modest amount of money to spend, you could afford to own and run a car. I remember as a student I could afford to buy and run a cheap second hand car, earning money with a Saturday job. Most of my student friends were the same.

Today that is no longer the case. Partially, this is due to the cost of car ownership: fuel, insurance and tax all cost much more than they used to, but there is a bigger underlying cultural change. The concept of freedom is changing, particularly in young people. In the past, a young person's first car was their first step to freedom. Everyone from my generation can remember their first car and how they used it to meet up with their friends. Today, many young people do not associate freedom with travel in the same way as their parents did. Instead, their friends are on Facebook, Skype, WhatsApp and SnapChat. They can play games together across the internet via mobile phones. When they meet together, they're more likely to play computer games or spend their time contacting a wider group of friends on their phones together rather than go out for a drive.

Consequently, fewer young people are aspiring to owning a car. Cars are often regarded as commodities rather than a treasured possession. The concept of ownership is changing as well. Young people have grown up with free stuff: they get a brand new mobile phone for free every two years with their phone contract; they listen to music on Spotify for free and they watch music videos and films on demand, also for free, on YouTube. They get their e-mail for free from Google or Hotmail.

Of course, a lot of these things aren't completely free: you get your phone for free because you've signed up for a two year contract, but the perception is often that the phone hasn't cost you anything. Young people ask why cars should be any different.

Unfortunately, that model doesn't work well for conventional cars. You pay for the car, you pay to service and maintain it, pay to fuel it, pay to insure it. The costs just keep stacking up. Yet an electric car can work a different way. With fuel costs virtually removed from the equation, an electric car can be sold in exactly the same way as a mobile phone: choose how far you wish to travel each month and get the car, the maintenance and the insurance for one fixed monthly cost. Even better, with the ever decreasing prices of electric cars, it is possible in many cases to get all that for less than the monthly cost of pumping fuel into a conventional car.

## Cost

When electric cars first appeared, they were expensive to buy. They still are, based on the sticker price, but scratch the surface and the prices are far more reasonable. If you are buying new, there are discounts to be had at your dealer. Leasing deals in particular are very good value for money, with cars available at very low monthly costs with low deposits. Want to buy second hand? Thanks to the popularity of the LEAF, supply is plentiful and prices compare with similar aged regular cars. The time when you had to pay a big premium for driving an electric car are over.

Once you've got your electric car your running costs are negligible. Once-a-year servicing and a slight increase in your household electricity bill are your only real costs.

If you can afford to own a car, you can definitely afford to own an electric car. If you are struggling to afford your own car at all, you should definitely consider an electric car.

# A brief history of the Nissan LEAF

Like many car manufacturers, Nissan had toyed with electric vehicles many times throughout its history. After World War II, petrol was scarce but electricity was plentiful in Japan. In 1947, the Tama electric car was launched. With a top speed of 35km/h (22mph) and a range of 96km (60 miles), the tiny four seat car found a market as a city taxi.

*The 1947 Tama electric car*

Beyond that, Nissan created a number of electric prototypes, concepts and limited production test vehicles at various times between the 1970s and the 1990s. Most had limited range and modest performance and were only hand-built in small numbers.

Electric vehicles remained an obscure niche until the mid-2000s. A growing awareness of the environment and the increasing costs of oil were bringing electric cars onto the agenda. The increasing demand for cars from emerging economies was also being felt. Whilst the majority of automakers stayed on the side lines or made vague announcements, Carlos Ghosn, CEO of both Nissan and Renault made a bold move. In 2007, Ghosn committed more than $5 billion into a development plan to create an entire line-up of electric cars under the Nissan and Renault brands. His rationale was simple; "If you are going to let developing countries have as many cars as they want – and they are going to have as many as they want one way or another – there is absolutely no alternative but to go for zero emissions. And the only zero emissions vehicle available today is electric. So we decided to go for it."

It was an unprecedented move, met with disbelief within the motor industry. As a senior executive at Ford told me at the time. "Ghosn has bet the future of two major car manufacturers (Nissan and Renault) on the viability of the electric car. The worldwide sales of electric cars this year (2007) is under five hundred vehicles." Other industry experts agreed. Ghosn was taking a huge risk. BusinessWeek magazine even branded Ghosn as 'crazy'.

The Nissan design team responsible for the electric vehicle products were starting from scratch. Unlike today, there was no company capable of mass-producing vehicle-spec lithium-ion batteries, the motor and control systems that existed were designed for industrial applications rather than automotive. The technology and products that did exist were being hand-built in tiny volumes and were incapable of being mass-produced without significant investment. It was a daunting challenge.

Nevertheless, the project began to take shape. A brand new battery design was produced and tested. New motors and control systems were developed. A number of prototype electric cars, based on the Nissan Cube and Nissan Versa were produced and road tested. I drove one of the latter on a short test run in the early 2009 and was greatly impressed with how easy it was to drive.

Meanwhile, the car design began to take shape. Rather than modify an existing vehicle architecture, the decision was made to create a brand new architecture specifically for the new electric car. The reason for this approach was simple: the motor and transmission for an electric car are far smaller than a conventional car, but the battery pack is both heavy and bulky. By creating a new vehicle structure rather than adapting an existing model, it would be possible to design the vehicle around its electric powertrain rather than create a compromise that would never be quite as good as a conventionally powered car.

*Right: Carlos Ghosn shocked the motor industry when he announced a $5 billion electric car development project back in 2007.*

Meanwhile, Ghosn was pushing the electric car message hard. The new car would feature in the competitive 'C-segment' class, targeting cars like the Ford Focus, the Volkswagen Golf and Toyota Prius. Nissan withdrew the Versa from the market: if you wanted a 'C-segment' car from Nissan, you had a choice of electric power or nothing at all. It was another bold move. Either Ghosn was crazy, or the world was about to see the next big thing in vehicle development. Either way, there was no doubting that Ghosn was a man of his convictions. The Nissan electric car project was going to have a lot to live up to.

The first Nissan LEAF was unveiled to the public for the first time on August 2nd, 2009 in Yokohama, Japan, and a few weeks later in the United Kingdom and the United States. Nissan announced the car would be in production and on sale by December 2010. The LEAF immediately caught the public's attention for its distinctive looks, its promise of near-silent driving and its all-electric powertrain. Time magazine declared it one of the fifty best inventions of the year and consumer response was enthusiastic: Nissan received over 20,000 pre-orders for the car in the United States alone and had to stop taking customer orders four months before the car officially went on sale.

### The EV-12 prototype

The EV-12 prototype was an advanced drivetrain mule produced by Nissan in order to carry out tests on the Nissan LEAF motor, batteries and control systems. These prototype vehicles were tested all around the world to ensure the car would work in all climates and conditions.

These grainy photographs of EV-12 were taken during its first public outing in May, 2009. The author is behind the wheel.

Production started on October 22nd, 2010 and on December 3rd the LEAF finally went on sale in Japan and the United States. Reviewers were impressed with the performance, the refinement and how easy it was to drive. It was noted that the car was extremely relaxing, was quieter than a Rolls Royce Phantom and that its instant torque meant that the LEAF could pull away from a standing start extremely rapidly, outpacing virtually any other car on the road in normal driving.

The car was packed with gadgets. A dashboard with twin digital displays. A keyless entry system. Reversing cameras, a digital warning sound for alerting pedestrians at low speeds and a general feeling of 'high tech' wizardry about the interior. On freezing cold mornings or baking hot days, LEAF owners could activate the heating or air conditioning remotely from their mobile phones or via the internet. Owners could also check the range of their cars from their phone or remotely switch on or switch off charging when the car was plugged in to take advantage of low-rate electricity.

The car quickly started winning major awards: first the 2010 Green Car Vision Award, followed by the 2011 European Car of the Year, the prestigious 2011 World Car of the Year and the 2011-2012 Car of the Year – Japan. The car captured the imagination of the mainstream press. Several early customers found themselves at the middle of a media frenzy with their car purchase being televised and owners appearing on television, radio and in newspapers and magazines. Owners were being stopped in the street by enthusiastic bystanders. Everyone wanted to know what they were like to drive, how long it took to charge and what the range was like.

Of course, not everybody was convinced. The LEAF was expensive, the styling was not to everybody's taste and there was a big concern about the range. Some consumers were worried about the possibility of running out of charge and being left stranded. This was new technology, how reliable was it going to be? How safe was it? How long would the batteries last? If the batteries died, how much would it cost to replace them? A lot of environmental detractors also questioned the claims of environmental efficiency, claiming, falsely, that the battery production negated any environmental benefits over running a conventional car and pointing to the fact that electricity is still being generated by coal-fired power stations and claiming, also falsely, that this made them even worse than a conventional car (the environmental impacts of using an electric car compared to a conventional one are covered in greater detail later in this book).

Gradually, the new owners started learning about their cars. They discovered the joys of walking out of their house in the mornings and finding the car fully charged, ready to go and never having to go to a service station to refuel. Many owners reported lower stress levels, thanks to smooth power and a lack of engine noise and vibration. Driving was good, too. There was instant power at the touch of the pedal, thanks to instant torque, no gears and the immediacy of electricity, which made inner city driving fun again.

All of the early cars were built in Japan, at Nissan's plant in Oppama, which could build up to 50,000 LEAFs a year. Production was heavily disrupted on March 11th, 2011 when a magnitude 9.0 undersea earthquake off the west coast of Japan triggered a series of powerful tsunami waves which hit the Japanese coast and travelled up to 10 miles inland. The earthquake was so powerful it shifted the Earth's axis by around eight inches. Despite being inland and over 200 miles away from the earthquake, the factory suffered structural damage and was without electrical power for several weeks. Limited production of the LEAF commenced at the end of March, gradually building back up to full volume later in the year.

Other car makers were watching progress closely. Was Nissan's gamble paying off, or was the LEAF going to be a sales flop? The truth, it seemed, was somewhere between the two. After the initial bust of public enthusiasm, sales of the car were steady, rather than phenomenal. People buying the LEAF tended to fall into two camps: families with more than one car, where a different car could be used for long distance journeys, and older people who were planning on owning their car for a long time and tended not to drive long distances. One thing that unified most customers was their desire to do something different, to own a new piece of technology. Environmental awareness was often part of the equation, but not always.

What held the LEAF back was the lack of a nationwide charging structure, particularly the rapid charging facilities that would allow owners to pull in, recharge their car in a few minutes and then continue on their journey. A few charging networks were emerging, allowing people to park up and charge up in towns and cities, but these mainly consisted of lower speed charging stations that took several hours to charge the car up.

Whilst the vast majority of electric car charging took place at home, or possibly at work, prospective owners wanted to know that they could charge their cars if they wanted to go further afield. Only then would an electric car be able to truly replace a conventional car in terms of flexibility.

Nissan started installing rapid chargers at their dealers around the country. Charging was available for free for Nissan owners. To stimulate the market, Nissan offered rapid charging equipment for free for charging network companies who were prepared to install and operate them. In the United Kingdom, Nissan teamed up with green electricity supplier, Ecotricity, who commenced a roll-out of rapid charging stations at motorway service stations and in central city areas. In the United States and Canada, where distances between cities is so much greater, building a nationwide network was far more daunting. Instead, rapid chargers were deployed for enhancing local travel rather than for driving from one major city to the next.

Progress was not all plain sailing. Some of the early rapid chargers were not particularly reliable and LEAF owners found they could drive to a charging station only to find that it was out of order. Many of the rapid charging stations, particularly at Nissan car dealerships, were only accessible during business hours and could not be used in the evenings or at night. The rapid chargers could only charge one car at a time, which sometimes meant a delay waiting for someone else to finish charging. Sometimes, the charger was blocked by an inconsiderate driver parking their car in the electric car reserved parking bays.

Gradually, the teething problems were ironed out. Chargers were upgraded and moved to more accessible locations where they could be accessed day and night. Multiple chargers were installed in popular locations. Additional charging points were installed in new locations. Clearer signage and fines for inconsiderate parking in electric car charging points dissuaded other car drivers from blocking charging points.

Nissan were also developing the LEAF technology further. The battery packs were updated and improved as the technology progressed. The LEAF itself

received a significant update in 2013, resulting in an updated car with a longer range, more interior space, a more efficient heating system and a faster home charging option that allowed the car to be recharged from flat to full in around 3½ hours.

The updated LEAF was now being built in multiple factories across the world. Cars for the North American market were being built in Nissan's factory in Smyrna in Tennessee. This factory also became the largest vehicle battery plant in the United States, capable of producing batteries for 200,000 vehicles a year. European LEAFs were being built at Nissan's Sunderland factory in the United Kingdom. Plans were put in place for building LEAFs in Guangdong, China. Not only did Nissan lead the electric car market, it now had the production facilities for producing far more electric cars and electric car batteries than any other manufacturer.

After initially being high, prices were falling. In Europe, Nissan were offering leasing deals where customers could buy the car but lease the battery, reducing the purchase price to the same as conventionally powered cars. Battery leasing was calculated to be cheaper than the cost of pumping fuel into a conventional car. Worldwide, Nissan were offering great car leasing deals too, leasing the whole car including batteries, at very competitive prices. When customers understood that they could lease a Nissan LEAF on a two or three year contract for a similar cost to their existing fuel bills, sales started growing fast.

The LEAF gained a bigger brother. The eNV-200 all-electric van brought electric power to the small commercial market. Available as a van, a taxi or a spacious five-seat estate car, the eNV-200 offers the benefits of electric drive to

*The Nissan electric eNV-200 is available as a van, taxi or 5 or 7 seat estate car.*

16

commercial operators. It has been enthusiastically received by small businesses, councils and major fleet operators, particularly those operating in urban environments. In March, 2015, Nissan announced the seven seat version of the vehicle, making it the only genuine all electric multi-purpose vehicle available today.

Sales have continued to grow by between 30-40% per year. The LEAF is now regularly the number one best-selling car in Norway. Worldwide sales reached 158,000 by the end of 2014.

Today, the Nissan LEAF is not the only family-sized electric car available. There are similar sized cars available from Volkswagen, Ford and KIA, amongst others. Yet the LEAF easily remains the best-selling electric car available today, with year-on-year sales still increasing rapidly. It is not hard to see why: the car has the best charging infrastructure available to it, it is the most cost effective to buy and it is built to a high standard. The decision to build the LEAF as a dedicated all-electric car instead of adapting an existing design has resulted in a car with best-in-class interior packaging, resulting in more space for people and luggage.

Eight years after Carlos Ghosn first made the announcement that Nissan were going to develop an electric car, Nissan has become the undisputed leader in this new, emerging market. It has unrivalled sales and worldwide production facilities and the best financed research and development teams in the industry. Sales volumes when compared to conventional cars are still small, but it is one of the fastest growing sectors of the market with the potential of completely changing the car buying market of the future.

Like it or not, the electric car has arrived.

# Living with a LEAF

## First impressions

The Nissan LEAF is a very distinctive car, unlike any other car on the road. Consequently, it always attracts attention. Despite the fact it has been around for several years, public interest in the car is still high. I am regularly asked questions about the car by interested bystanders. I had my car on charge at a public charging point in my local town recently. When I came back, a group of tourists were taking photographs!

People watch with curiosity as the car pulls away virtually silently. There is no getting away from it, if you buy a Nissan LEAF you will have to get used to the fact that you are going to attract attention.

Sit inside the car and there is no mistaking the fact you are sitting in a car with plenty of gadgets. There is a twin-screen digital dashboard that can feel a little intimidating when you first sit inside the car, but is clearly laid out and easy to learn. Early cars all had light cream interior, with a light grey and cream

dashboard and seats, giving the car a light and airy feel. Since 2013, the cars have had a darker interior that feels less airy, but enhances the high-tech feel of the cabin and is easier to keep clean. There are some neat details, such as the domed gear lever, gloss black trim for the central console and metallic blue trim around the switchgear. The impression is of a solidly built, quality interior.

The seats are very comfortable and there are plenty of adjustments available. Rear seat passengers have plenty of headroom although on the earlier cars (before 2013) the rear floor level is slightly higher to accommodate the battery pack. Later cars have improved this, making it more comfortable for very tall passengers.

Of course, the big difference with the Nissan LEAF over a conventional car is what happens when you press the START button. Instead of the cough of an engine being fired up, the car plays a short chime, the dashboard lights up and then... silence. Pulling away for the first time is an unusual sensation. The lack of noise and slight vibration that you get with an internal combustion engine makes it feel slightly unreal. Slide the gearshift into drive, press the accelerator and the car just pulls away. Press the accelerator harder and the car accelerates rapidly, catching most people by surprise. Most people do not expect an electric car to be quick, but the LEAF certainly is, particularly when accelerating from low speeds.

It takes a few minutes to get used to the sensation of travelling without engine noise. At speeds of up to 15mph, the car makes very little noise at all. The LEAF has an electronic 'engine' sound that plays through a speaker in the front of the car to alert pedestrians when the vehicle is travelling at low speeds, but this is virtually inaudible from inside the car. As the speeds increase, the car is still very quiet, with the electric motor sounding like a muted jet turbine.

## Performance

On paper, the Nissan LEAF should not be a quick car. With 108bhp and a heavy body, the car shouldn't feel quick. The official performance figures of 0-60mph in 11½ seconds and a top speed of 89mph does not look impressive on paper, yet everyone who drives a Nissan LEAF comments on how good the performance is.

There are three reasons why this is. Firstly, when you push the accelerator pedal in a conventional car, the extra fuel has to be pumped from the fuel tank to the engine. The engine then has to burn this fuel and then you get the power. It all happens in a fraction of a second, but it is still a delay. In an electric car, the surge is absolutely instant. The power is delivered to the wheels with an immediacy that is very welcome, but takes a little getting used to.

The second reason is down to the characteristics of an electric motor compared to a conventional engine. Electric motors may not be powerful, but they do provide a lot of torque. In fact, the Nissan LEAF's motor has a similar amount of torque to a 2.5l V6 engine. More importantly, whereas you have to rev up an engine to deliver the torque, an electric motor gives maximum torque from a standing start. Consequently, the Nissan LEAF is one of the quickest family cars on the road when pulling away from the traffic lights.

The third reason that a Nissan LEAF feels quick is that there is no gearbox. The electric motor drives the wheels through a single speed transmission. If you need to accelerate hard, there is no need to change gear (or for the car to do change gear itself if you drive an automatic). Consequently, there is no delay before the car accelerates.

And finally, it is worth pointing out that whilst many family cars have faster official 0-60mph times, few can genuinely deliver that performance in everyday driving. Achieving the official 0-60mph times in a conventional car means high revs, a racing start, red lining the engine and a racing gear change. Few drivers have the skills to do this and the additional wear and tear on the vehicle transmission is likely to result in some expensive repair bills if it is done too often. Accelerating quickly in a Nissan LEAF is far simpler: if you want to achieve the official 0-60mph time, just press the accelerator.

Most people when pulling out onto a fast road will take between 18-30 seconds to accelerate from zero to 60mph. Even if you have a more powerful sports car, you really have to be trying to out accelerate a Nissan LEAF from a standing start.

The result is a car that feels very nippy with lots of usable power that is very easy to drive. Because of the instant response, the car can pull out quickly into a gap in the traffic and rapidly accelerate to match the speed of the cars around you. If you are travelling at higher speeds and need to accelerate quickly to overtake somebody, the car's instant power makes it easy and safe to do so.

So whilst the figures may not look impressive, the reality is that powering a car with an electric motor is a step change in the evolution of the car. Jeremy Clarkson, former presenter of Top Gear in the United Kingdom, summed up the difference between an electric car and a conventional car when he said "there must have been a moment in history when everybody had typewriters and typewriters had been around for hundreds of years and they were brilliant... and then someone came along with a laptop..."

## Braking

In a conventional car, the braking system slows down a car using friction. Brake pads press against the brake disk, creating friction and slowing down the car. All the energy used to accelerate the car is lost to heat.

The Nissan LEAF, in common with most electric cars, has two braking systems. As well as the conventional mechanical braking system, the car has an electronic *regenerative* braking system to slow the car down. Regenerative braking uses the momentum of the car to generate power to put back into the batteries. In effect, the motor becomes a generator to recharge the car.

When you take the foot off the accelerator, the motor starts gently recharging the batteries using the momentum of the car, without significantly reducing the speed. Touch the brake pedal gently and the regenerative braking level increases, putting more energy back into the batteries. If you press the brake pedal harder, the mechanical brakes also cut in to supplement the regenerative braking.

With the Nissan LEAF, you can adjust the amount of regenerative braking using the gear lever. In *DRIVE* mode, the LEAF behaves much the same as an automatic car with very little regenerative braking. In *BOOST* mode, the regenerative braking effect is much greater, slowing the car down more rapidly when you take your foot off the accelerator.

Regenerative braking is a great way of recouping some of the energy that would otherwise be lost. With practice, it is possible in boost mode to carry out virtually all braking purely using regenerative braking. Only if you need to brake in a hurry or bring the car to a complete halt will you need to use the mechanical brakes. The difference regenerative braking makes to range is significant, particularly in city driving where learning how to get the best out of regenerative braking can increase the range of the car by up to thirty percent.

## Handling and Ride

The Nissan LEAF was never designed to be a serious driver's car. Suspension is set up for comfort rather than for sporty driving. Despite that, the car is well balanced and with the batteries mounted low down beneath the floor, the car remains composed around the corners, even when taken at speed. The steering is nicely weighted to make the car feel secure however fast you travel.

If you prefer to relax and enjoy the smooth ride, the LEAF does a good job of absorbing the bumps. It makes it a great car for commuting and carrying out the daily duties of a family, whilst offering enough fun for more spirited driving.

## Refinement

The LEAF is one of the quietest cars on the planet, comparable to the Rolls Royce Phantom. The electric motor is virtually silent at lower speeds and emits a muted jet turbine-like sound at higher speeds. There is no vibrations from the engine and the smooth power delivery makes for comfortable driving.

In a conventional car, the sound of the engine masks other noises, such as the noise of windscreen wipers, the rumble of the wheels on the road and the wind noise as the car slices through the air. Nissan has had to work hard to reduce or mask these noises in their electric vehicles so that they do not detract from the travelling experience.

The result is a fantastically quiet car. You can drive through the countryside in the spring and summer with the windows open and listen to the birds singing as you cruise past at 50mph. Although there is no official research on electric cars and stress levels, many LEAF owners I know say that this level of refinement has resulted in them being far less stressed when driving, particularly when commuting in rush-hour traffic.

## The Fun Factor

Most people who have never driven an electric car are quite surprised by how much fun they can be. The Nissan LEAF is no exception. The characteristics of the motor make it a nippy car to drive. The car is heavier than most in its class because of the weight of the batteries. However, as the batteries are mounted low in the chassis at mid-point in the car, the car remains composed when driven hard.

Watching other people's reaction can be fun, too. Bystanders are often surprised when the car pulls away without any engine noise. Passengers also like the lack of noise and are often impressed with the performance.

## CARWINGS®

CARWINGS is an on-board computer system with a live data link through to Nissan. It provides traffic and travel information and ensures the latest electric vehicle charging points are recorded on the built-in satellite navigation system.

One of the real benefits of CARWINGS is that it allows you to access your car via the internet or from a smartphone app and check on the charge level, or switch on and off the heating and air conditioning. Third party apps have been developed that also allow you to set the heating onto a timer to switch on in advance.

Consequently, you can turn on the heating in winter, or air conditioning in summer, and get to the car a few minutes later when it is at the right temperature. Not only is this a great comfort feature, but it allows you to heat or cool the car whilst it is plugged into a charging point, therefore not impacting on the range of the car when you get in.

CARWINGS also has the ability to send you an e-mail if the LEAF is not plugged in by a certain time in the evenings, or if charging stops part way through. If you are worried that you'll forget to plug in your Nissan LEAF overnight, it can remind you.

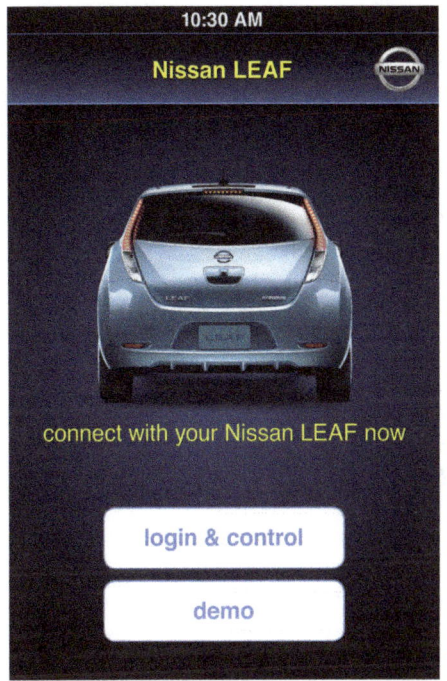 

*The Nissan CARWINGS® mobile phone app allows you to check your charge capacity and turn on the heating or air conditioning.*

# Range

You cannot discuss any electric car for very long before the discussion focuses on the range. All around the world, it is the number one concern that *non*-owners have about owning one.

It's an understandable concern. With a conventional car, you take it to a service station to refuel it, then drive it around for 3-400 miles before pumping in more fuel again. There are plenty of places to go to refuel and it takes only a few minutes.

There is a different principle for an electric car. Whenever you leave your car at home, you plug it in. When you get back to it, it's usually fully charged and ready to go wherever you want. Some people are able to charge their cars when they are at work. The same principle applies, by the time you get back to the car, it's usually full charged and ready to go.

In fact, research suggests that between 85 – 95% of all electric car recharging is carried out at home. Owners rarely need to charge anywhere else because the car has enough range for most of the journeys they do. Public charging points, either the rapid charging that recharges the car in minutes or the lower

speed charging points dotted around towns and cities, accounts for between 5 – 10% of electric car recharges.

Consequently, drive an electric car for a few days and you will find you are thinking about range and fuel in a different way. Many electric car owners will describe the freedom they feel every time they go out to their cars in the morning: the car is fully charged up and ready to go. They know they have enough range to go wherever they want to, without the hassle and cost of visiting a service station to refuel.

*Between 85-95% of all charging is carried out at home. A dedicated charging point can be installed inside a garage, or on an outside wall and can the car up from flat to full in as little as four hours.*

One electric car owner explains it like this. "It takes me nine seconds to charge up my electric car. That is the time it takes to plug the car in when I get home. The next time I need to use the car, it is charged up and ready to go." Several drivers talk about how convenient it is to plug their car in at home each night for a full charge the next day, as opposed to the inconvenience of driving to a service station to refuel every few days.

If you drive around 12,000 miles a year, your average daily driving is around 33 miles a day. The average American car owner travels 29 miles per day by car, with an average single journey of 12 miles. In the UK, the average British car owner travels 22 miles per day by car, with an average single journey of 6½ miles. 93% of all car journeys in Europe are shorter than 25 miles.

Official European figures for the original LEAF was 109 miles (175km), whilst the US figures recorded an official range of 73 miles. After the LEAF was updated in 2013, the range increased, with official European figures of 124 miles (200km). The US figures show an official range of 84 miles.

The reason for the difference in official figures is down to different testing processes. The *New European Driving Cycle* tests carried out for all cars are carried out in a laboratory at a fixed temperature, simulating a completely flat route. Wind is removed from the equation. Unsurprisingly, the resulting figures are optimistic to the extreme and provide an unrealistic guide for consumers wishing to know the true range of an electric car.

The United States has adopted a far better model for range, with a series of tests designed by the Environmental Protection Agency (EPA) to give consumers a much clearer idea of the efficiency of their car and the range that they could expect to achieve. Their tests use a combination of city driving and highway driving cycles, whilst using a certain amount of air conditioning or heating where required. These figures mean that the ranges quoted in the US are far more realistic than those quoted in Europe. Many LEAF owners exceed the official US range figures with their own driving, depending on how and where they use their cars.

Incidentally, it is worth noting that a new worldwide set of test procedures is currently being developed which will replace both the NEDC and EPA tests. Called the 'Worldwide Harmonized Light Vehicles Test Procedures' (WLTP), these tests are designed to reflect real-world driving far more accurately than either the NEDC or EPA tests, and will start being used in 2016.

In real life, both reviewers and owners have reported very different results on real-world range. Batteries are affected by ambient temperature, so you get better performance in the summer than the winter. Driving styles and speeds make a big difference, whilst running the heating or air conditioning also has an impact, particularly on the earlier cars.

However, there is only one way to ascertain real-world range, and that is to try it yourself. As part of the research for this book, I carried out a number of test journeys in order to ascertain the realistic range in two Nissan LEAFs: a 2012 model that had previously done 21,000 miles and a 2014 car that had previously done 4,000 miles.

A series of tests was carried out using two different drivers to ascertain the 'real-world' range of the LEAFs. The cars were both driven on public roads in an ambient temperature of 17–20°C (62–68°F), without the use of heating or air conditioning. The tests were driven from cars with a full battery, stopping when the range indicator went blank and the 'extremely low battery' warning message was shown. At this point, the car has a reserve range of between 3 and 5 miles.

## Real world range testing

| | 2012 LEAF | 2014 LEAF |
|---|---|---|
| Official range (European NEDC tests) | 109 miles | 124 miles |
| Official range (US EPA tests) | 73 miles | 84 miles |
| City driving | 90 miles | 109 miles |
| Urban/extra urban | 81 miles | 97 miles |
| Cross country roads | 75 miles | 82 miles |
| Cross country roads – fast driving | 46 miles | 58 miles |
| Dual carriageway/freeway eco driving | 65 miles | 74 miles |
| Dual carriageway/freeway normal driving | 51 miles | 58 miles |

The tests were then repeated in the winter, in temperatures of between -2 – 4°C (28 - 39°F), using heating when needed to keep the cabin at a comfortable temperature:

| WINTER TESTING | 2012 LEAF | 2014 LEAF |
|---|---|---|
| Official range (European NEDC tests) | 109 miles | 124 miles |
| Official range (US EPA tests) | 73 miles | 84 miles |
| City driving | 81 miles | 94 miles |
| Urban/extra urban | 71 miles | 89 miles |
| Cross country roads | 67 miles | 76 miles |
| Cross country roads – fast driving | 39 miles | 51 miles |
| Dual carriageway/freeway eco driving | 55 miles | 69 miles |
| Dual carriageway/freeway normal driving | 40 miles | 51 miles |

As you can see, both the newer LEAF and the older LEAF show a drop in the colder temperatures. The drop is more significant with the 2012 model, as the older car has a less-efficient heating and air conditioning system. When in use, the older system could reduce range by as much as 25%. The enhanced heat-exchanger system that Nissan released in 2013 is far more efficient and has much less impact on range.

## What happens when a LEAF runs out of charge?

*The Nissan LEAF instrument panel showing:*

A: Speedometer          B: Charge Remaining (percent)
C: Fuel Gauge           D: Estimated Range (miles)

The instrument panel shows how much charge is left in the car with a fuel gauge (C), a range estimator (D) and – on the new cars – a battery percentage gauge (B). When the car gets down to around 17% charge, the low battery charge light illuminates and the message 'Battery Level is Low' is shown on the vehicle information display. If your car has satellite navigation, this will show a warning message, offering to navigate you to your nearest charging point and the voice message 'Low Battery' is given. You typically have a range of around 15 miles at this point.

If you continue driving, a second warning message is shown. The driving range on the instrument panel begins to flash and the voice message 'Extremely Low Battery' is given. At this point, you have a range of between six and eight miles and the driving range display (D) changes to "- - -".

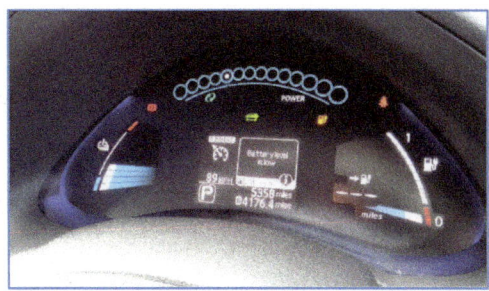

*As the range drops, an orange fuel sign lights up and the estimated range display starts flashing. When the battery charge range drops to under 8%, the estimated range changes to three dashed lines.*

If you continue to drive further, the car goes into 'tortoise mode'. A picture of a tortoise appears on the instrument panel and the motor power is significantly reduced. At this point, you have around half a mile to pull over safely and stop.

## How much is enough range?

A lot of people who are attracted to the cost benefits of driving electric are longer distance drivers. If you find you are a high mileage driver, the idea of saving 90% of your fuel costs is an attractive one.

In the United Kingdom, the average Nissan LEAF owner does drive higher than average miles. There are many owners, including me, who are driving around 20,000 miles a year with their cars. Some of the early cars are now reaching 100,000 miles and still performing extremely well.

If your high mileage comes from lots of shorter trips, with the opportunity to recharge between journeys, then a Nissan LEAF can work very well. If your high mileage comes from longer distance journeys, then you are going to be reliant on rapid charging along your route.

Longer distance journeys are feasible if you live in an area with a good rapid charging network (such as the United Kingdom and some parts of the United States), but if you need to go long distance and use the rapid charging network more than a few times a month, it soon becomes wearying.

Of course, if you have more than one car in your family, a range of more than 50 miles is rarely an issue. Going long distance on a business trip? Arrange to swap cars with your partner for the day. Going on a family trip? Likewise, take the other car.

Remember, too, that the LEAF can be charged between road trips. I regularly drive 50 – 60 miles in the morning, plug my car in for a couple of hours, and then drive a similar distance in the afternoon, so if you do lots of journeys throughout the day but can recharge between them, your *daily* range is significantly increased.

## Range Fixation

The first time you get into a Nissan LEAF, you will be very aware of the fuel gauge. In fact, if you have never driven an electric car before, the fuel gauge becomes a fixation for the first few weeks. An electric car has a shorter range than a conventional car with a fuel tank, so the fuel gauge moves more quickly than you will first expect.

In most conventional cars the fuel gauge seems to stay close to full for the first half tank of fuel and then moves down quite rapidly afterwards. In any car, once the fuel gauge drops below a quarter, many people start getting 'range fixation'. They are on the lookout for a fuel station and start to worry if they cannot find one on their route.

In an electric car, the fuel gauge will start to move after only a few miles of driving. It is a bit disconcerting at first, because everyone is so used to the way fuel gauges work in conventional cars. In effect, you are getting the same range fixation as you do when running low on fuel in any other car. Even when you are driving a short distance and you absolutely *know* there is enough charge to get to your destination, it is very easy to get fixated on range in the early days.

Alongside the fuel gauge, the Nissan LEAF also shows the estimated range remaining. This can fluctuate quite dramatically depending on how you are driving: drive at 25-30mph in a built up area and the range is high. Go onto a faster road and travel at 60-70mph, or drive up a steep hill, and the range drops alarmingly quickly. The reality is, of course, that most journeys are made up of a combination of faster and slower roads, so the estimated range remaining figure can be a little bit of a misnomer. But for the uninitiated LEAF driver, it can be alarming and increase the worry about range.

It is a psychological difference. In reality, you are no more likely to run out of range in an electric car than you are to run out of fuel in a normal car. The best analogy to use is that of a mobile phone. If you plug your phone in overnight to charge it up, it has enough charge to last the next day. The same is true with an electric car. Once you have more confidence in your electric car and have used it for a while, your range fixation disappears. In fact, you get to the point where you ignore the fuel gauge completely. After all, if you plug your car in every night and you know that you have enough range to do your daily driving, why bother checking the fuel gauge?

## The first few weeks

The first few weeks with a Nissan LEAF are fun. The novelty factor of a car that runs on electricity lasts a while and the driving characteristics of an electric motor makes driving very entertaining.

Some people go through a period where they are thinking about every journey you are taking, constantly checking that they have enough range. When going out of town, they might even make arrangements to plug their car into a power socket at their destination, even if their destination is well within the range of the car. Thankfully, this soon wears off as people get more confident about driving their cars.

Owners are understandably nervous when taking their first long distance drive, where they are then reliant on either a rapid charge on route or on charging their car at a public charging point at their destination. The nerves are caused by unfamiliarity: once you've done it once, you know what you are doing and you can relax a bit more about recharging when you are out and about.

One thing that happens to most people at some point in the first few weeks of electric car ownership is that they forget to plug the car in overnight. Most people do it once – usually at the most inconvenient time! The Nissan LEAF CARWINGS system can be set up to remind you if the car is not plugged in by a certain time in the evening, thereby reducing the risk of you forgetting. If you do forget to plug in, it is usually not the end of the world. If your journey is relatively short, you can often plug your car in and get enough range after 20–30 minutes of charging to get you to your destination. As a consequence, you might be late but the results are rarely catastrophic. However, most electric car owners rarely make the same mistake twice!

## Long term ownership

As you settle into every day ownership, you'll stop thinking about range, plugging in the car and so on. It will simply be second nature. You may find that your driving style becomes more relaxed as you adapt to get the best out of the regenerative braking system. Conversely, you may also find that you enjoy being the first away from the traffic lights as the car pushes you back in your seat with the initial burst of acceleration.

Generally, LEAF owners love the convenience of having a car with 'a full tank of electricity' every morning. They love the smoothness and refinement of the electric motor and they love how easy their car is to drive.

Another thing that most Nissan LEAF owners do agree on: once they've made the switch to electric power they do not want to make the switch back to a conventional car again. Several owners, myself included, are already on their second LEAF. Driving an electric car is seen as an improvement and a genuine step forward.

# Plugging it in

The biggest difference between an electric car and a conventional car is how you refuel it. As I mentioned in the previous chapter, the average electric car owner carries out around 95% of their charging from home, with the remaining 5% of charging either carried out at public charging points or at the home of a friend or family member.

## Charging from home

You can plug a Nissan LEAF into a domestic power socket, but as this is likely to be your primary charging location, this is not recommended.

It takes a long time to charge using a domestic power socket. If you are based in Europe, a domestic power socket provides up to thirteen amps of power at 220/240 volts. Charging a completely flat battery takes around eleven hours. In the United States, a domestic power socket provides up to fifteen amps of power, but only at 120 volts. Charging a completely flat battery takes fifteen hours.

For many owners this is good enough. A full charge may take a very long time, but if you only drive twenty or thirty miles in a day, your car can be fully recharged in three or four hours.

In North America, charging at 110 or 120 volts from a domestic power socket is often referred to as *level one* charging or *trickle charging*. It is not the fastest way to charge, but it is certainly the most available.

Most electric car owners choose to install a dedicated charging point for their car. Technically called *Electric Vehicle Supply Equipment* or EVSE, these charging points work at 240 volts at either 16 or 32 amps and are sometimes referred to as *level two* chargers.

These charge the Nissan LEAF far quicker. A 16 amp charging point can provide 3.3kWh of energy and can charge a Nissan LEAF in under eight hours. The 32 amp charging point can provide 6.6kWh of energy and can charge a Nissan LEAF from flat to full in around four.

It's worth noting that prior to the 2013 upgrade, the LEAF could only charge up at 16 amps. The latest LEAF can charge at 32 amps, and therefore in half the time, although this is an extra-cost option on some models.

A charging point can easily be installed at your home by a qualified electrician. You can choose to have it fitted inside a garage, on an outside wall of your home or, in some cases, on a post on your driveway.

The costs vary quite dramatically and it is worth shopping around. In the United Kingdom, as of April 2015, a Government grant is available to cut the cost of the charging point and the

*This home charging point comes complete with cable. Unplug from the wall, plug into the car and you're charging within seconds.*

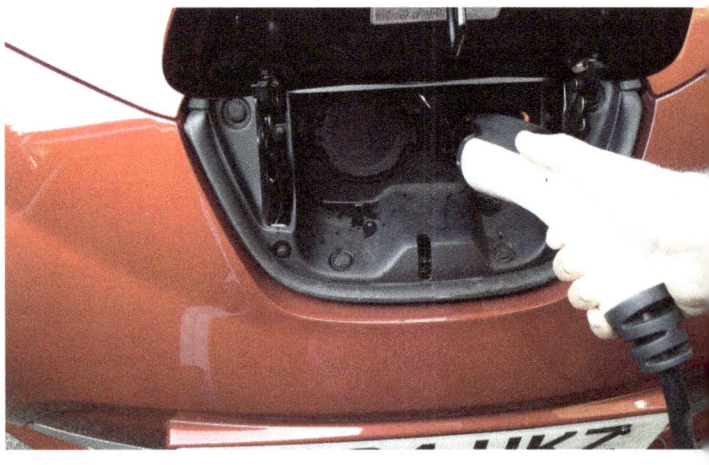

installation by up to 75%. Many installers are offering free charging points as part of this scheme, providing a 25% cashback on completion.

If you have to buy a home charging point, prices can vary from around $300 up to $1,800 (£200 to £1,000), plus installation. Some charging points incorporate the charging cable, which makes it more convenient for plugging in. Others simply have a socket and you have to plug your charging cable into the charge point.

When you buy your LEAF, Nissan provide you with a cable that allows you to charge your car up from a domestic socket as standard. A different cable is required for plugging your car into a dedicated charging point. Depending on the specification of the car you are buying, Nissan may provide you with this cable as part of the purchase price of the car. Otherwise, you will need to buy one unless you are purchasing a charging point with built in charging cable.

## What to look for when buying a home charging point

### Before you begin

Usually, most houses will be able to have a charging point installed without over-stressing the household electrics. However, before buying a charging point, it is worth getting your household wiring assessed by a qualified electrician.

This is particularly important with older houses. If you have an older house with aging electrics, your electrician may advise that you only install a low capacity charging point.

### Location

When installing a charging point, you have a choice of where to install it. You can install it inside a garage or car port, you can install it on an outside wall of your house, or you can install it on a post on your driveway.

The most expensive of these options is installing it on a post. Running underground electrics to the post is time consuming and expensive and the post has to be installed on proper footings. This requires additional construction work which can quite significantly bump up the cost of installing a charging point in your home.

### Capacity

Home charging points come with a choice of capacity: the two main options are 240v 16 amp/3.3kW and 240v 32 amp/6.6kW.

If you have a Nissan LEAF that has the 3.3kW charging capacity, either charging point will work for you. You can still plug a 3.3kW LEAF into the higher capacity 32 amp charging point, but there will be no difference in charging time. You can then choose whether to save money and buy the lower capacity charging unit or 'future proof' your system by opting for the 32 amp version.

If you have a Nissan LEAF with the 6.6kW charging capacity, you will only get the benefits of the faster charging with a 32 amp charging point. You can still plug your car into a 16 amp charging point, but your charging time will then increase.

## Tethered Cable or Socket

There is a saying in the computer industry: 'the great thing about standards is that there are so many to choose from'. Regrettably, the same is true for electric car charging standards.

There are two different standards for the charge plug on an electric car. Nissan, Mitsubishi, Kia, Renault, Citroen, Peugeot and Ford, amongst others, opted for the one plug (the J1772). BMW, GM, Mercedes, Smart and Volkswagen opted for another (the 62196-2).

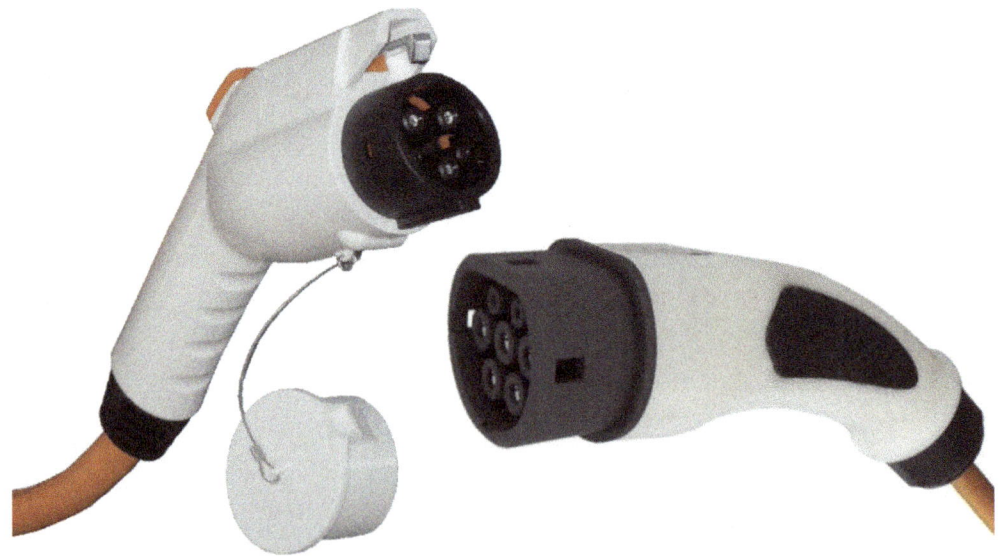

*Left: The J1772 plug as used in the Nissan LEAF.*
*Right: the 62196-2 plug used in some other electric cars.*

Thankfully, whichever electric car and whichever plug you have, you can charge them all using a public electric car charging point. From a real-world owner's point of view, the only real difference is the physical shape of the plug on the car itself.

However, if you are buying a charging point for your home, you have to choose whether you want a charging point with a tethered cable, or with a socket that you then plug your charge cable into.

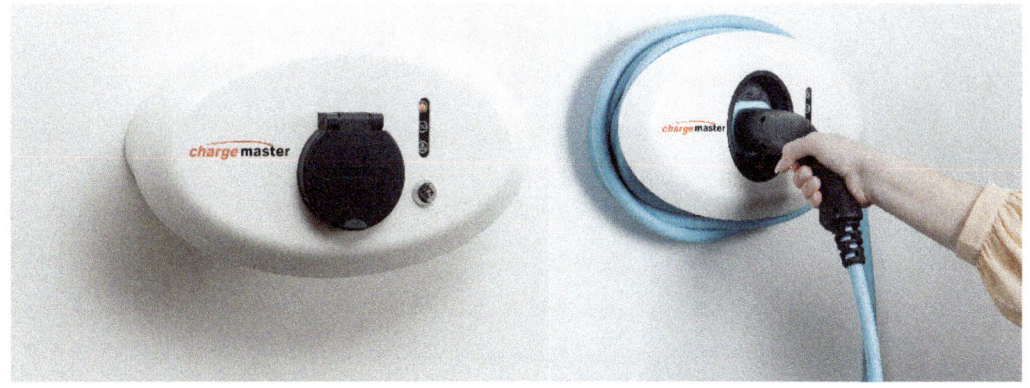

*Left: an example of a home charging point with a socket connection.*
*Right: a home charging point with a tethered power cable.*
*Pictures courtesy of Chargemaster Plc.*

The benefit of buying a charging point with a tethered cable is convenience: you don't have to get your charge cable out of the car and unravel it. However, the downside is that your charge point will only work with other electric cars that have the same charge socket.

The benefit of buying a charging point with a socket is that this is universal: it will work with all makes and models of electric car.

## Home charging point suppliers and installers

Below is a list of companies who can supply and install electric car charging points in your region. This is not an extensive list. In the United Kingdom, for example, you can buy electric car charging points over the counter at a number of electrical distributors, whilst a number of companies supply charging points through the Amazon website in the United States and Canada.

**Australia**

| | |
|---|---|
| Chargepoint | www.chargepoint.com.au |
| Origin Energy | www.originenergy.com.au |

**Canada**

| | |
|---|---|
| Bosch | www.pluginnow.com/residential |
| Chargepoint | www.chargepoint.com |

**Ireland**

| | |
|---|---|
| Baird Electrical (NI) | www.bairdelectrical.co.uk |
| ESB | www.esb.ie/electric-cars |
| Carra | www.carra.ie |

**NZ**

| | |
|---|---|
| Chargepoint | www.chargepoint.com.au |
| Origin Energy | www.originenergy.com.au |

**UK**

| | |
|---|---|
| Chargemaster | www.chargemasterplc.com |
| Pod Point | www.pod-point.com |
| Rolec | www.rolecserv.com |

**USA**

| | |
|---|---|
| AeroVironment | evsolutions.avinc.com |
| Bosch | www.pluginnow.com/residential |
| Chargepoint | www.chargepoint.com |
| Clipper Creek | www.clippercreek.com |

## Charging at work

Many businesses are now providing electric car charging in their company car parks. Some owners have been able to persuade their employers to either provide a dedicated charging facility or allow them to charge up from a standard power socket.

There are several options for employers considering offering charging points for their employees. Many companies are considering electric vehicles for their own use, either as pool cars or with electric commercial vehicles such as the Nissan e-NV200 panel van, and providing extra charging facilities for employees can be done at relatively low cost. Companies can do this to demonstrate their own eco-friendly credentials, but also to ensure staff flexibility: if an electric car can be charged up on site, it becomes easier for that employee to undergo unplanned business travel if required at short notice.

At its simplest form, work charging can simply be conventional weatherproof external sockets installed on an outside wall for electric car use. If faster charging is required, or additional security so that it is not easy for the electricity supply to be misappropriated, dedicated charging points can be installed. These can vary from a domestic-type charging point for one or two vehicles, to a commercial version complete with tag readers which will only unlock when a pre-programmed ID card is presented to the charge point. Charging points can either be mounted to an outside wall or provided as a dedicated stand-alone charging post.

Costs for a fully installed system for one or two cars range from around £500 to £2,500 in the United Kingdom and between $750 and $3,500 in the United States. Instead of outright purchase, many companies instead choose to lease a system, which can work out to be far more affordable. In the United Kingdom, you can lease a system from around £45 per month.

## Public charging points

The past five years has seen a huge explosion in the number of public electric vehicle charging points being installed. Virtually every city and major town across Europe now has a network of charging points, and whilst North America is not as advanced, it is catching up quickly.

Many countries also have rapid charging points on major roads. In the United Kingdom, for example, virtually every motorway service station and many 'A' road services now have rapid charging points installed that allow LEAF owners to recharge their cars in around 30 minutes. Ireland, too, has a full nationwide charging point infrastructure

Public charging points broadly fall into two categories: destination charging and rapid charging.

*Public charging points, such as this on-street charging post, are now*
*a common sight in major towns and cities. This destination charge*
*point can charge two cars at the same time.*

Destination charging uses level two chargers to charge your car slowly over several hours. They are designed to be used to either provide a small top-up, or to fully charge your car whilst you leave the car for a few hours. Destination chargers are typically installed in cities and at shopping malls. They are ideal for people working in towns or for longer shopping trips when you are expecting to be away from the car for several hours.

Charging using a level two charging point takes the same time as charging from home using a dedicated charging point. If you have a newer LEAF with the 6.6kW charging system, that means a full recharge from completely flat to completely full will take around four hours, whilst if you have the LEAF with the 3.3kW charging system, a full recharge will take around eight hours.

## Rapid Charging

Rapid charging – sometimes referred to as DC charging – uses a level three charging point. Level three chargers are high power chargers using a three-phase industrial power supply. This is converted to a high voltage DC current and used to charge the Nissan LEAF batteries very rapidly. How rapidly depends on various factors such as ambient temperature, the age and condition of the batteries and how much energy is already stored in the batteries, but typically a Nissan LEAF can be recharged from flat to 80% full in around 30 minutes. If you leave the car plugged into a rapid charger for longer, the car will continue to charge up, but at a much lower rate. Consequently, most people who use a rapid charger to recharge their cars tend to unplug once the car gets to 80% charge in order to continue their journey.

Unfortunately, there are a few different standards for rapid charging, which are all incompatible with each other. The CHAdeMO standard used by Nissan, Mitsubishi, Peugeot, Citroen and KIA is the standard in most widespread use. CHAdeMO is used in the LEAF. The other mainstream standard, CCS, is being rolled out rapidly. In the main this means that existing rapid charging stations are being upgraded to support both CHAdeMO and CCS charging in the same unit.

*Ecotricity's network of rapid charging points across the UK's motorway network has been largely funded by Nissan.*

Consequently, you will find that most rapid charging points either offers CHAdeMO charging and nothing else, or offers both CHAdeMO and CCS. At present, there are very few rapid charging points that only offer CCS.

There is also a proprietary 'supercharger' charging station used by Tesla in their Model S and Model X cars. This standard is only supported by Tesla. The limited number of supercharger charging points are all owned and operated by Tesla for use by their owners.

Because CHAdeMO rapid charging is the closest to being universal, you are less likely to have problems finding somewhere to rapid charge your LEAF, compared to owners of other electric cars that do not use the CHAdeMO standard. Combine this with the fact that many Nissan dealers also have CHAdeMO rapid chargers that are freely available to LEAF owners and you have a much better chance of finding rapid charging points with a LEAF than any other electric car.

However, if you are looking for a rapid charging point for the first time, it is worth making sure that the charging point you are heading for is a CHAdeMO charger. Thankfully, this is fairly straightforward with a number of websites providing a worldwide database of charging points. The website I tend to use is www.openchargemap.org. If you limit your search to CHAdeMO charging points, this website will allow you to easily identify the rapid charging points in your area.

## Charging Point Network Operators

There are a number of different companies offering public charging points. A few years ago, electric car owners had to sign up to various different schemes in order to ensure they had access to all the different networks, but new cross-company agreements now mean that owners need only join one or two networks to cover an entire country.

Most charging companies offer a choice of different schemes, starting from a simple 'pay as you go' scheme where you buy credits in advance then top them up as you need to, through to a monthly scheme that gives you access to a number of recharges each month. If you are lucky enough to live in the United Kingdom, where electric car charging is helped by significant Government support, you can subscribe to networks for a very small fee and then have unlimited charging within your region at no additional cost.

When you subscribe to a charging point network, you're given a credit-card sized ID card. You hold this over the card reader on the charging point, which then identifies you and allows you to plug in your car. With destination charging, it is usual to have to supply your own cable for charging whilst with rapid charging, the charging station itself has a cable that you plug into the car. The cable that you will require in order to charge up at most destination charging points is the same cable used for the dedicated home charging points.

*You will require a dedicated charging cable to charge up from public charging points, or for charging your LEAF from a home charging point fitted with a socket rather than a tethered cable.*

The following is a list of charging point network operators for different regions. This is not an exhaustive list and other network operators are available.

If you intend to use public charging points, you will need to register with a local charging operator. They will then supply you with a credit-card sized access card that you can use to access the charging points.

| Country | Operator | Payment Options | Comments | Web Site |
|---|---|---|---|---|
| Australia | Chargepoint | Pre-pay membership or pay-as-you-go via credit card | Destination and rapid charging, based in Brisbane, Sydney, Victoria, Adelaide and Perth. | chargepoint.com.au |
| Canada | Chargepoint | Pre-pay membership or pay-as-you-go via credit card or phone app | Destination and rapid charging in major cities. | chargepoint.com |
| Ireland | ESB | Free | Nationwide network of destination and rapid chargers. Owners have to request a free access card. | www.esb.ie /electric-cars |
| NZ | Chargepoint | Pre-pay membership or pay-as-you-go via credit card | Destination charging only for Aukland and Wellington. | chargepoint.com.au |
| UK | Charge Your Car | Pre-pay membership | Access to virtually all destination chargers in the United Kingdom | chargeyourcar.org.uk |
| UK | Ecotricity | Free | Nationwide network of rapid charging points at motorways and dual carriageways. Owners must sign up to receive a free access card. | www.ecotricity.co.uk /for-the-road |
| USA | Chargepoint | Pre-pay membership or pay-as-you-go via credit card or phone app | Destination and rapid charging in major cities, with a growing number of rapid chargers situated on major roads. | chargepoint.com |
| USA | EZ-Charge | Free/pre-pay | Multi-network charging scheme, includes the Nissan 'No Charge to Charge' scheme for free charging. | ez-charge.com |

# Putting your car on charge at a Destination Charger

Step 1: Use the in-built satellite navigation to identify your nearest charging point

Step 2: Pull up at the charging point. Hold your Charge ID card over the card reader on the charging point. Follow the on-screen instructions.

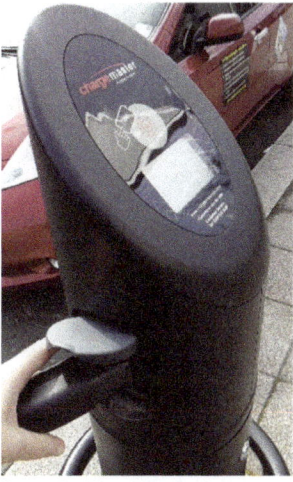

Step 3: Plug your charge cable into your car. Then plug the other end into the charger.

Once you have plugged your LEAF in, it beeps twice to acknowledge that it is plugged in and on charge. The cable is locked into the charge point until your return.

## Unplugging your car on your return

Step 1: Hold your ID card over the card reader on the charging point. After a few seconds, you will be prompted to remove your cable.

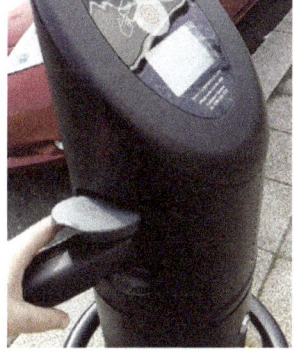

Step 2: Unplug your cable from the charging point.

Step 3: Unlock your car, disconnect the cable from the car and close the charge flap.

# Charging up at a Rapid charger

Step 1: Hold your ID card over the card reader on the charging point. Follow the instructions on the screen.

Step 2: Some charging stations have multiple charge options. If asked for which charger to use, choose 'DC' or 'CHAdeMO'.

Step 3: Connect the charge plug into the rapid charge socket on the LEAF. Squeeze the trigger on the plug to lock it in place.

Once you have plugged your LEAF in, check the charging station to see if there are any further steps to follow in order to start the charge as some charging stations ask you to confirm the car is plugged in before initializing charging. The LEAF beeps twice to acknowledge that it is on charge.

## Unplugging your car on your return

Step 1: If your car is still on charge, press the CANCEL CHARGE button and follow the instructions on the screen.

Step 2: When prompted to disconnect, squeeze and release the trigger on the plug in order to unlock it. You can now unplug the cable.

Step 3: Replace the charging cable into the charging station. The charging station is ready for use by the next LEAF owner.

# Free Electric Car Charging

There are free electric car charging schemes in the United States, the United Kingdom and Ireland.

## United States

For new LEAF owners, Nissan now offers a two-year 'No Charge to Charge' scheme, giving access to a network of public charging points for free. The scheme allows up to 30 minutes of free-charging at rapid chargers and 60 minutes of free charging at 'level two' destination chargers.

The scheme is administered by EZ-Charge (www.EZ-Charge.com). Details of the No Charge to Charge scheme will be provided by your Nissan dealer when you purchase your LEAF.

Nissan are also in the process of rolling out a further 500 rapid charging stations at Nissan dealers throughout the United States. These are also part of the No Charge to Charge scheme.

## United Kingdom

If you live in the United Kingdom or Ireland, many of the charging points can be accessed entirely for free, thanks to a combination of government grants and funding from car manufacturers, particularly Nissan.

In the United Kingdom, you can currently access all of the Ecotricity charging points entirely for free. You simply visit their website and register for an access card, then use this card to access the charging points on your route. Ecotricity rapid charging points cover virtually every motorway service station in the UK, along with service stations on many dual carriageways. They also have rapid charging points in several city locations. Ecotricity have stated that they have no plans to start charging for this network in the foreseeable future.

Nissan also have rapid charging at many of their dealerships. These are available to all Nissan LEAF owners to recharge their cars entirely free of charge. Nissan have stated that these charging points will always be available to Nissan LEAF owners free of charge.

## Ireland

Ireland possibly has the most comprehensive network of charging points in the world, with a large network of both rapid chargers and destination charging points covering major routes and towns and cities across the country. The network is implemented and managed by ESB and details can be found at www.esb.ie/electric-cars.

The scheme covers both the Republic of Ireland and Northern Ireland, with drivers able to cross the border and charge seamlessly across the country.

At the time of writing, access to the entire network is free, thanks to a significant amount of European Union funding. This will change at some point, at which point the charging points will be accessed by a pre-pay card system.

50

# Will a LEAF work for me?

Before making the leap and buying an electric car, it is important to consider whether it is a practical proposition for you. Range is a real problem for a few people and a perceived issue for many more. There are answers to these issues, but you need to make sure that you are either happy to accept that on occasion, you may be compromised by an electric car, or happy that you will not have these problems in the first place.

## What benefits do I expect to get from owning a LEAF?

There are many benefits in owning a LEAF. From competitive leasing costs and low running costs, to the environmental benefits, the convenience of charging from home and the ease of use.

You need to ask yourself, however, what is important to you. For example:

- Low running costs
- Better fuel economy
- The opportunity to drive something different, new and interesting
- Reducing your environmental impact

## Where do I live and where do I drive?

If you live in a town or city and most of your driving is in built-up areas, a LEAF makes a lot of sense. It is non-polluting, it is easy to drive and there are usually lots of places you can charge it up whilst you are out and about.

If you live in a small country village miles from anywhere, the LEAF still makes sense, particularly if there is no service station nearby for putting fuel into your current car. You'll probably be more reliant on your car and travel longer distances than a 'townie', and are more likely to be caught out by unplanned journeys. For this reason, you should consider the LEAF with the 6.6kW home charging option that can be recharged relatively quickly at home: a four hour recharge time means that you are less likely to be caught out with a flat battery than an electric car with an eight hour recharge time.

If you do a lot of travelling on high-speed roads for long distances, you will need to check that the distances you are travelling are comfortably within reach of the LEAF. As shown on page 29, the range of the LEAF drops when travelling at higher speeds or when driving in colder conditions. Realistically, if you are travelling more than fifty miles per charge on a daily basis, a LEAF is probably not a practical proposition for you right now.

## How far do I travel in an average day?

Unless you have actually measured how many miles you travel on a daily and weekly basis, you are probably not aware of all the different journeys you do in your car and how far you drive.

People who claim to know how many miles they travel each day are often surprised when they actually measure their journeys, as opposed to estimating them. Most people over-estimate their journey lengths by between a quarter and a third and forget to take into account one fifth of all their car journeys.

Keep a diary for a couple of weeks, noting when you use your car and how far you travel each time you use it. Also, make a note of the type of roads you are driving on and your average and top speeds. The results could surprise you and will give you accurate figures of what you really need your car to do for you.

Once you know how far you travel on a daily basis, you will now have a clear idea of what you need to achieve with a LEAF. Then compare the range you require with the range figures shown on page 29 and see how they measure up.

## How often do I travel long distances?

If you need to travel longer distances, then you are going to need to consider public charging, either with rapid charging so you can extend your journey, or with destination charging so you can recharge before your return journey.

If you are based in the United Kingdom or Ireland, there is a nationwide network of both rapid charging and destination charging points already in place. You can go virtually anywhere you want to go, with charging points at motorway service stations and major dual carriageways, as well as in many towns and cities. In these countries, it is possible to travel many hundreds of miles in a day using a Nissan LEAF.

*High speed charging points, like these two, have become a common sight at motorway service stations in the United Kingdom.*

It is also possible to use your car away from home. I've taken my Nissan LEAF on a number of family holidays, travelling upwards of 150 miles to get to my destination, then reliant on public charging points whilst on holiday to charge my car. It has worked very effectively.

If you are based elsewhere, you will need to plan your journeys a little more carefully. You can look up the location of charging points online, using websites such as openchargemap.org, to identify what charging facilities exist along your route and plan accordingly.

Practically, however, long distance journeys have to be planned. Whilst it is possible in the UK to travel hundreds of miles in a single day, the practicalities of this makes it less than ideal. Recharging at a rapid charger takes between 30-45 minutes, and whilst this may be acceptable once on a journey, it becomes tiresome if you have to do it three or four times in a single day. I have driven almost 300 miles in a single day in a Nissan LEAF. I can't recommend it as a practical way of travelling on a regular basis.

An alternative to this is to hire a car for the odd occasions that you need to travel further afield. This can be a practical and very cost effective solution, and allows you to choose a suitable car for the purpose: a large car for travelling with a group of people, a small car for another occasion, even a luxury car for impressing your boss. A full days car rental through companies like Enterprise can cost as little as £25 in the United Kingdom or $30 in the United States.

The third alternative for long distance journeys is public transport. Travelling by train can be faster and more comfortable than taking the car, and prices need not be high, particularly if booking in advance or travelling during off-peak periods.

## I do not have off-road parking. Can I still own LEAF?

Practically, you really do need somewhere at home where you can park and charge up your electric car. However, that has not stopped some people coming up with ingenious solutions to owning an electric car, even if they do not have off-road parking.

In the United Kingdom, there is a government scheme that will pay for the installation of a charging bollard and allocating a reserved parking space for residents that do not have off-road parking. A growing number of councils have subscribed to this scheme.

Elsewhere, electric car owners have arranged for the installation of street-side charging bollards themselves. Other enterprising electric car owners have arranged for a cable run to be installed into the pathway so they can safely plug in at night.

If this sounds like an option, go and talk to your local council and see what they say. If you do get approval, there will be a number of conditions: you will need to use a council approved contractor for installing the charging point and they may levy an additional charge on you for permission. Most of the charging point manufacturers and installers have experience of installing on-street charging points. They may also be able to help you work through the paperwork and ensure that everything is done correctly.

As an alternative, some owners have been able to persuade local businesses to allow them to park and charge their cars on their premises overnight. This can be a mutually beneficial arrangement, as it means there is activity at the premises during the evening, which in turn reduces the risks of the premises being broken into.

## What if an electric car is not suitable for me now?

This is a good point to take stock. If you have already decided against buying a Nissan LEAF, then that is a shame, but at least you have vital information. It is better to have read this book and decided against buying the LEAF than to have spent thousands on a LEAF and then discovering that it is not suitable for you.

Electric cars are continuing to evolve and improve. Even if a LEAF is not a practical solution for you now, it may well be in two or three years' time.

There is another option that you may wish to consider: car sharing.

If you live close to friends or family who would also be interested in owning a Nissan LEAF, why not pool resources and buy or lease one between you? If you then share all the cars you have, you can use the LEAF for shorter journeys and another car for long-distance driving.

There have been instances where communities have clubbed together to buy and run an electric car. The car is parked in a permanent location such as a local community area or library, where the car can be kept on charge. The keys are held in a wall-mounted safe with combination lock, and booking is set up using one of the many freely available internet-based diary systems. When somebody in the scheme wants to use the electric car, they simply book it, pick up the keys and drive away.

There are other ways of sharing a car, too. In the UK, Easy Car Club (www.EasyCarClub.com) allows you to register your car for car sharing, so you can earn money from your car while it is not in use. There are similar schemes in the United States, such as Getaround (www.getaround.com) and Relay Rides (www.relayrides.com).

Sharing a car may sound radical, but it is gaining momentum, with schemes in many major cities. In big cities such as New York, Chicago and London, car sharing is now commonplace, with companies such as Zipcar, Car2Go and Hertz all offering car-sharing schemes. In the UK, over 140,000 people have now subscribed to car-sharing clubs.

# Me and my LEAF

LEAF owners come from all walks of life. Some of them own a LEAF because of the environmental benefits, others because they want to drive something different or because they are attracted by the low running costs of their cars.

With the exception of this introduction, this entire chapter is written by other people. These people already use a Nissan LEAF as their day-to-day vehicle. This is their story.

## Damian Powell, Birmingham, UK.

### 2013 Nissan LEAF

My old car was becoming expensive. Every service required unexpected work on top of the work that I knew about, the fuel economy was getting poorer, the mod-cons were, well, not modern, and it didn't have any cup holders. So I decided it was time to get a new car, and by new I mean brand new, for the first time ever. I've always been interested in the idea of green motoring and after going for a ride in a friend's LEAF I decided that now was the time to try.

My initial impression of the LEAF was that it is a very sharp car to drive. That is, it feels very tight and very responsive while at the same time being relaxed and comfortable. I can honestly say that I enjoy driving this car. Unfortunately for me my partner drives the car more than I do because I work from home and she has a relatively long commute; it seemed silly for me to have the more economical car at home while she spends money on a relatively inefficient petrol driven car. The upside is that I get to drive it at the weekend and I look forward to doing so.

What would I say are the major pros of owning an EV? My trips to service stations are now far less frequent. The only time I go there is to check wiper fluid, or to get fuel for me. The car itself is very comfortable, well specified and intelligently designed. I also think that the car has helped me to become a better driver. This

make it easy to understand that bad driving is often inefficient driving. Since driving the LEAF, I find that I drive my ICE car more considerately.

There is one major drawback with an EV. It's the obvious one: range. Now I'm not saying that range is poor on the LEAF. In fact, the range of the LEAF is quite good and I very rarely need to use charging-points away from my home. However, the range can be severely impacted by adverse conditions. Wind, rain, cool temperatures, high speeds, and steep inclines – these can all be detrimental. The real problem is not that the range is affected, but that it is very difficult to predict how badly it will be affected.

The other major con is the availability of charging-points. It's fair to say that the availability of charging-points has probably doubled in the year since I have had my LEAF. That's certainly true when talking about rapid chargers. Other drivers are inconsiderate of EVs, though, and often park in spots reserved for EV use.

I love the LEAF. I love driving it. I love the fact that I'm not having a detrimental effect upon the environment when I drive it (I get my electricity from a green energy supplier). However, the infrastructure is still not there to make it a carefree driving experience (although it has nearly doubled in just the last year) and legislation is not yet there to treat rare, EV charging spaces with the same reverence as disabled parking spaces. I think these things will continue to change over the next couple of years though, and as battery technology improves and ranges increase, I expect that EV driving will become normal. That is to say, we won't have to pay special attention to the weather, we won't have to drive slower than all of the other vehicles, and we won't have to worry about being caught short because there'll be rapid chargers at all the major service stations. My next car will also be an EV.

## Peter Berg, California, USA.

### 2011 Nissan LEAF

I have always been interested in green technology. In 2002, I bought my first new car ever, a Toyota Prius hybrid. It was a beautiful cobalt blue and was the most efficient car I had ever driven. At the time, it seemed that the Prius was the most advanced and efficient vehicle that was available.

I was not aware of any electric vehicle options until I read an article about the new Aptera EV that was being developed. My interest peaked, I paid the deposit and got on their waiting list. I was bitten by the electric vehicle bug and I wanted to be a part of the revolution I knew was coming.

The turning point for me was when Nissan announced their new LEAF electric car. They hosted an event here in Los Angeles and invited those interested in being early adopters. At the event, I saw a prototype of the car, learned about how their car would work and saw their battery technology. It was then that I knew I wanted an electric vehicle. I got on Nissan's list and paid the reservation fee right away.

In December of 2010, after eight months on the waiting list, I received an email telling me that our car had been produced in Japan and was on its way to California. What a wonderful early Christmas present! We got a call from our dealer a few weeks later and picked up our all-electric car in January 2011.

Our dealer was sixty miles away in Fontana, California. Danny, the sales rep, called on a Tuesday to let us know our Cayenne Red LEAF was ready. Did we want to pick it up tonight or wait until the weekend to make the drive out there? My choice was obvious. We went to the dealer after work and arrived a little after 9 pm. When we got there, the staff were as thrilled as we were, despite the late hour. The manager stayed late to make sure she shook our hands, and photos were taken. We were some of the first people in the United States to get this exciting new car.

The car comes with great features: a built-in GPS navigation system, Bluetooth technology to connect to my phone, XM satellite radio, live traffic information, a USB port to play music with, and more. Nissan did a great job with the engineering, and the build of the car is sleek and modern. It has definitely gotten some attention, both on the road and off.

Normally I only drive about 30 miles per day so the car works great for my daily work commute. We have been able to take it on a few longer trips as well. We went to visit my wife's parents in Orange County, and it was an easy trip there and back. We also took a trip up into the hills to go for a weekend hike, and it handled the hills and curves with ease.

Keeping the car charged has been equally trouble-free. After one of those longer trips, the car can be charged in about six to seven hours, but with my short commute I can get away with plugging in only two or three times a week. We might have been able to use only our main 120 volt power source to charge the car, but I thought the faster 240 volt option would be good when we need a faster charge. I found a company that made an EVSE (electric vehicle supply equipment, or charging dock) that worked with the LEAF and I installed it myself. With the solar panels on our roof charging the car from the sun, we now are able to lower our power bill *and* our carbon footprint.

Owning the LEAF has made me think a little bit differently about driving now. I have to do a small amount of planning to make sure we will have enough charge for our trip and to decide which vehicle to take (we still have our Prius). While some people say that doing that planning is too much for them or too inconvenient, I have found that it is not a big deal. It is certainly worth the advantages that come with the car: no more trips to the gas station, much less maintenance with the car, a smoother and quieter ride, lower operating costs with the car, the cutting-edge technology that comes with the car, and, most importantly, less pollution going into our air.

There have been a few issues with the car, but nothing major. I did have to get used to the charging settings. You can set up timers to delay the charging until the evening but I had to learn how the charging timers worked. The first night, my car did not start charging since it was after the start time on the timer. I used the override switch to solve the problem for the first night. I modified the timer and now it is not a problem.

Despite these few small issues, we love the car. It is easily the most fun car I have had. While I wanted a fun car, our main reason for getting the car was to make the air we breathe a little cleaner. Here in Los Angeles our air quality is not good. It is often hazy and unhealthy to breathe. It is obvious that the air quality is a result of man's activities, so it is only fair for us to step up now to do more to reduce the pollution going into the air and water. The secondary reasons to get the car were the lower operational and maintenance costs, the rebates that made the cost of the car very reasonable, and the smoother ride that the car gives us.

I think the LEAF is a great option for many people. Anyone interested in getting an electric vehicle should at least test one out. If you see how far you really drive in a day you might be surprised how easy it would be to start driving an electric vehicle. The car keeps you informed about how far you can go, so you do not have to worry so much. Range in an electric vehicle will be no more a concern than in a traditional gas car once more charging-stations are installed in public areas. Just do not go out on long cross-country trips: rent a car for those instead of putting the wear and tear on your own car.

Having a clean and green electric car is a great choice, and I hope a lot more people get rid of the gas-guzzlers and start driving one of these fun and efficient electric cars. They really are a great way to go.

## Diane Courtney, London, UK.

### 2014 Nissan LEAF

This is my second Nissan LEAF. We bought the first one in September 2011, intending it to be used just for my commute to work and for daily family duties, carting the kids to their clubs and doing the weekly shop.

The LEAF soon proved to be a far more versatile vehicle and it quickly became our main family car. If we wanted to go somewhere as a family, we would always take the LEAF rather than our gas guzzler. My husband and I would end up fighting over the keys for the LEAF, simply because it was so much fun to use. The kids far prefer to be driven around in the electric car and most of their friends still think that the LEAF is cool.

When I got my first LEAF, I was quite anxious about range to start with. I was very conscious that the car could only do around 80-100 miles per charge, without ever really thinking about how far I travelled each day. Having got used to a car that could drive 350 miles or more on a single tank of fuel, I did worry that I would not have enough range to do the driving I wanted to do. I was also quite disconcerted by how much the range dropped when I switched on the heater. Switching the heating on could make the range drop by around 15 miles.

After a few weeks of driving the LEAF, however, I decided that I was worrying over nothing. The fuel gauge rarely went below half-full during my normal daily driving, even if I had the heater blowing full blast throughout my journey.

Initially, we were only planning to use the LEAF for shorter journeys, but as more charging points were installed, I began to use it for longer distances as well. Charging from a rapid charging point takes between 30-45 minutes, for some reason it seems to take longer in the winter than the summer, but is generally quite convenient. The only criticism is that the rapid charging cable itself is very bulky and the charge connector can be a bit of a fiddle. Only once have I had to wait a few minutes at a charging point because somebody else was charging, but now multiple rapid chargers are being installed which should resolve that problem in the future.

I changed jobs around two years ago and ended up having to travel far further for work, increasing my commuting to around 50 miles a day. This never really caused any problems with range as I would plug the car in when I got home. If I needed to use the car for short journeys in the evening, there was always enough charge. On the odd occasion when I was going to be using my car for a longer journey in the evenings, I would use a rapid charger and give my car a ten to fifteen minute charge boost on my way back from work whilst I took the opportunity for a loo break.

I had intended to keep my LEAF for longer, but unfortunately the car was written off in a traffic accident last October. When we considered our options for a replacement vehicle, we carefully considered our options. Should we get another LEAF, or choose an electric car from Volkswagen, BMW or Ford? We did not even consider a non-electric car.

In the end, the choice of another Nissan LEAF was the obvious one. Nissan have updated the car significantly since our original car and the new one has a better range, which is far less affected by switching on the heating compared to the original car. I can also charge up at home in half the time. In practical terms, this means I can get home in the evening after travelling 50 miles to work and back, plug the car in for an hour and have around 80% charge available for any other journeys I want to do.

I drive around 1,600 miles each month. I cannot say I've noticed any big difference in my electricity bill in the years that I have owned my two LEAFs. I estimate that I am saving around £210-£230 a month in fuel compared to driving a small, economical diesel car.

About six weeks after we took delivery of the LEAF, our other car, an aging Mercedes 4x4 that had been living on borrowed time for at least five years, finally gave up the struggle and went to the big Scrapyard in the Sky. The Mercedes had ended up being used less and less over the years we had the LEAF and had got to the point of being almost redundant. We took the decision not to replace it and just rely on the LEAF as our only family car.

It's been a good decision, saving us money and not making any real difference to our day-to-day routine. The only regular long trip we do is to visit my parents once every month or so, who live 130 miles away. Thanks to the nationwide charging network, there are no shortage of places I can stop to recharge the car along the way. We tend to stop once, after driving for around an hour, and have a half hour break to stretch our legs and have a drink. We then get back to the car refreshed and ready to complete our journey with the car fully charged and ready to go. When we get to my parents, we then plug in and charge up from a domestic power socket ready for the return journey.

I love my LEAF. It has been a revelation and I never intend to go back to a gas-guzzling car again. I find it suitable for both short and long journeys, my children – now teenagers – both love it and it is a lot of fun to drive. I would wholeheartedly recommend it as a family car.

# Model Range

As with all cars, there are a number of different models within the LEAF range. All Nissan LEAFs are based on the same battery and motor and the differences between the different models are largely down to variations on trim, accessories and charging options.

The LEAF received significant updates in 2013. The improved car was lighter, had an improved battery system, 15% more luggage space, improved frontal aerodynamics and a far more efficient heating system. The result was a car with a significantly better range and improved charging facilities, including the ability to charge your car at home in half the time.

## North America

When the LEAF was launched at the end of 2010, it was available in two versions: the SV and the top-of-the-range SL. The SV included satellite navigation and internet/smartphone connectivity to the vehicle for monitoring charge levels and switching on heating or air conditioning. The SL model added a number of features such as reversing camera, solar panel spoiler and automatic headlights and windshield wipers. On 2011 cars, the rapid charge facility was an optional extra across the range, but this was incorporated as standard on the SL model in 2012. There were minor variations of specification in different states, with features such as the cold weather pack standard in the north of the country but optional extras in the south.

When the car was updated in 2013, a third model, the entry-level S, was added to the line-up and prices across the whole range were reduced.

As with most cars, specifications and exact prices vary between states and are subject to minor variations from one year to the next, but at time of writing (March 2015), the specifications for the three versions are as follows:

## LEAF S

- Keyless door entry and ignition with push button stop-start.
- Bluetooth hands-free phone system
- Heated front and rear seats

The LEAF S has restricted charging options. It does not include the rapid charge port for high speed on route recharges, nor does it have the faster 6.6kW charger for faster home and destination charging.

## LEAF SV

The LEAF SV has all the features of the S, plus:

- Touch screen navigation system
- 6.6kW on-board charger for faster home and destination charging (flat to full in under four hours)
- CARWINGS
- Heat pump air conditioning

The LEAF SV does not include the rapid charge port as standard, although it can be specified as an optional extra. It does, however, have the 6.6kW charger so that you can fully charge your LEAF in around four hours from a suitable charging point. CARWINGS gives you the opportunity to remotely check the charge levels and switch on the air conditioning and heating via a mobile phone app or an internet connection. It also ensures the car has the latest traffic and travel information and an up-to-date list of public charging points in your area.

The heat pump air conditioning system is far more efficient than the standard system, ensuring the range does not drop as significantly when cooling or heating the car.

## LEAF SL

The LEAF SL has all the features of the SV, plus:

- Quick charge port for rapid charging
- Leather seats
- Automatic on/off LED low beam headlights

# Europe

When the LEAF was launched in Europe, it was only available in one model. This changed when the car was updated in 2013, with four main models now available: the Visia, the Visia +, the Acenta and the top-of-the-range Tekna.

The early cars came with a very high level of specification, with satellite navigation, keyless door opening and ignition, cruise control, reversing camera and electronic air conditioning as standard. It is very similar to the specification of the newer Acenta. All of the early cars came with the facility to rapid charge the cars in around half an hour using the dedicated rapid charging point network that Nissan was putting in place.

The rapid charging facility became an optional extra on the bottom of the range Visia and Visia + models in 2013. This limited the benefit of the car and for 2015, Nissan reintroduced the rapid charging facility across all models of LEAF. If you are purchasing a used LEAF, it is worth checking that the car does have the rapid charge socket if you require this option.

One significant change in the 2015 line-up is that the 6.6kW charger that allows the LEAF to be charged up from a home charger or destination charger in under four hours, has been made an optional extra across the whole range of models. Instead, all cars come with the 3.3kW charger as standard.

The specification of the different models does vary from time to time, but here are the main differences between the different models at time of writing (April 2015) for the UK market:

## LEAF Visia

- Keyless door entry and ignition with push button stop-start.
- Bluetooth hands-free phone system
- Manual air conditioning
- Rapid charge socket

The current model of the LEAF Visia has the ability to be rapid charged within minutes (this was optional in 2014 models). It does not have the faster 6.6kW charger, however, so charging from home or from destination charging points takes longer.

## LEAF Visia +

The Visia + has all the features of the Visia, plus:

- Alloy wheels
- Rear view reversing camera

Like the Visia, the Visia + does not have the faster 6.6kW onboard charger. This is available as an extra-cost option on both the Visia and Visia+ models.

## LEAF Acenta

The Acenta has all the features of the Visia +, plus:

- Fully automatic air conditioning with pollen filter
- Satellite navigation
- CARWINGS®
- 'Boost' mode to increase regenerative braking
- Heat pump air conditioning and heating

CARWINGS gives you the opportunity to remotely check the charge levels and switch on the air conditioning and heating via a mobile phone app or an internet connection. It also ensures the car has the latest traffic and travel information and an up-to-date list of public charging points in your area.

The heat pump air conditioning system is far more efficient than the standard system, ensuring the range does not drop as significantly when cooling or heating the car.

## LEAF Tekna

The Tekna has all the features of the Acenta, plus:

- Leather Interior
- LED front headlamps with auto-levelling
- Heated steering wheel  and front seats
- BOSEX® sound system

# The Options list

The exact specification for each model changes from one country to the next, and will change from one model year and the next.

All models of the Nissan LEAF come with a fairly high specification as standard and of course, it is your choice if you wish to upgrade to a higher specification. Like all cars, you also have the opportunities to add personal touches to your car with various cosmetic enhancements, plus practical options such as carpet mats and the alike. However, from an electric vehicle perspective, there a few options that you may wish to consider.

## 6.6kW charger

One of the most useful options is the 32 amp/6.6kW charger that became available in 2013. This charger can half the time it takes to charge a LEAF at home, from around eight hours with a 16 amp/3.6kW charger down to around 3½ – 4 hours.

This faster charge facility comes as standard on some models but is an optional extra on others. In the UK, the 6.6kW charger came as standard on the Acenta and Tekna models in 2014, but has now been made into an optional extra for 2015. In the United States the 6.6kW charge option currently comes as standard with the SL and SV specification cars (as of April, 2015).

### How useful is this feature?

*The 6.6kW charger makes it possible to recharge relatively quickly between journeys. If you are planning to use your LEAF for lots of journeys throughout the day, or if you are planning to drive long distance journeys reasonably often, the 6.6kW charger is a worthwhile addition.*

## Rapid charging

If you are planning longer distance journeys, then the rapid charging option is a real benefit. Fitted as standard to the LEAF SL in the United States and now standard in all models in the United Kingdom, the rapid charging option makes it possible to charge up from flat to 80% full in around 30-40 minutes.

Of course, this is only a benefit if there are rapid chargers in your area. The UK and Ireland have a great network, some parts of the United States and Canada have plenty of rapid chargers, but in other regions there are very few. It is relatively easy to find out where charging points are online as there are a number of online databases that provide details of where the charging points are. The website I use can be found at www.OpenChargeMap.org.

As mentioned on page 43 on my section on Rapid Charging, the Nissan LEAF uses the CHAdeMO rapid-charging standard. When checking on an online database for rapid chargers, make sure you limit your search to CHAdeMO chargers in order to identify your nearest charging point.

### How useful is this feature?
*Rapid charging makes it possible for the LEAF to be an only car for a lot of people, but only if you live in an area where there are sufficient rapid charging points.*

*The ability to take the LEAF on longer journeys and charge up on route makes the LEAF a very practical car. You would not want to be stopping for rapid charging every day, but if you only travel longer distances and need rapid charging facilities a few times each month, this feature makes the LEAF a more practical proposition.*

## Satellite Navigation
The Nissan LEAF satellite navigation system is a top-of-the-range system with links for real time traffic and travel information. It also incorporates a number of enhancements specifically tailored to the needs of the LEAF owner.

For instance, the system incorporates a database of all local charging points, both 'level 2' (standard speed) and 'level 3' (rapid) charging facilities. It allows drivers to make the choice between taking the fastest route, the shortest route or the most eco-friendly route, so owners can choose between getting to their destination quickly or getting the best range out of their LEAF. If your journey is outside the range of the car, the satellite navigation can also look up charging facilities along your route and incorporate those into your journey.

### How useful is this feature?

*If you are planning to use your LEAF as your only car or will use it for longer journeys, the Nissan satellite navigation option is well worth having. Because you can see how many miles you have remaining on your journey, it gives you the peace of mind knowing that you have sufficient range to complete your journey. If you need to charge up on your route, the system can navigate you to charging points along the way.*

## CARWINGS

The Nissan CARWINGS system works with the satellite navigation and is responsible for providing the car with the up-to-date information for traffic and the charging points. It also allows you to monitor the car remotely, either using a mobile phone app or a website.

Using the website or the phone app, you can then switch the heating or air conditioning on and off, set up a timer to switch on charging at a certain time, or just check how much charge you have in the car, along with a range estimate.

### How useful is this feature?

*CARWINGS is a luxury, but one that I personally use extensively. In the cold months, switching on the heating before I get to the car is a real comfort, whilst the ability to check the charge status on my mobile phone is a real benefit, particularly if I am using a public charging point.*

## Heating and Air Conditioning options

Owners of the early Nissan LEAF will tell you that when you turn on the heating or air conditioning, the range of the car dropped quite substantially. This is because the heater was all-electric, therefore increasing the drain on the batteries.

In the 2013 update, Nissan launched an updated heating and air conditioning system that used a heat pump to create the right ambient temperature inside the vehicle. Heat pumps heat or cool the cabin using the temperature difference between a refrigerant and the outside air, thereby creating heat (or cooling) using far less electricity.

The heat pump is an excellent system and makes a big difference if you are using the car for long distance journeys. Unfortunately, due to its expense, it is not offered across the whole LEAF range as standard. In the United States, the heat pump is offered on the SV and SL models whilst in the United Kingdom, the heat pump comes as standard with the Acenta and Tekna models.

### How useful is this feature?

*For longer journeys, the improved heating system makes it possible to travel further without having to worry that the heating system is draining the battery excessively. If you are only intending to use your LEAF for shorter journeys, it is not such a big advantage.*

## Heated front seats

Another heating option on the LEAF is the heated front seats and steering wheel package. Using a fraction of the energy to heat the whole cabin, the heated seats and steering wheel often provide enough heating for the car in cooler conditions, with a negligible effect on range. The heated front seat and steering wheel comes as standard across all new LEAFs in the United States, but is only offered on the top-of-the-range LEAF Tekna in the United Kingdom.

### How useful is this feature?

*I have not bought this on either of my LEAFs, but people who have used them say that it does improve comfort, particularly on longer journeys.*

# Purchasing and Running Costs

If you are planning to buy your Nissan LEAF outright, you will find the purchase price is higher than many comparable internal combustion engine cars, although the price differential is far less than it was a few years ago.

Nissan have worked hard to reduce this difference in price. In Europe, they are offering a battery rental scheme which reduces the purchase price of the LEAF considerably. The battery lease is then priced to be similar or lower than the equivalent fuel costs in a conventional car.

There are also financial incentives for buying an electric car. In the United Kingdom, a grant scheme takes £5,000 off the price of the car. In the United States, there is a $7,500 Federal Tax Credit, plus additional state incentive schemes available. Unlike the British scheme, the tax credit is a rebate, so you have to pay the full price for the car and then claim the discount later.

## Leasing

Nissan offer very competitive leasing schemes for their cars, making this an extremely attractive option. Whilst many people would prefer to own their cars outright, leasing electric cars can make a lot of sense.

With a conventional car, one of the biggest ongoing costs is fuel. With an electric car, the cost of fuel is negligible. This means that electric car owners can save a significant amount of money each month because they no longer have to pay for fuel.

So what if you would love to own an electric car, but can't afford the purchase price? With Nissan's leasing packages, it is possible to lease a car for two, three or four years and pay a similar amount for the car that you would otherwise pay for fuel. In effect, instead of paying the oil companies your money so that you can run your existing gas-guzzling car, you're paying Nissan for a brand new car and getting rid of your fuel bill almost completely.

## Buying Used

Based on the US Kelly 'Blue Book' and UK's Glass's Guide, used Nissan LEAFs retain 33–38% of their new value at three years old, based on an average mileage and condition. This makes them an excellent buy as used cars.

Exact values do vary, depending on where the cars are based. In London, for example, where electric cars are exempt from the daily congestion charge, used prices are around 8% higher than the rest of the country. The same is true in California, where used electric cars are sold at higher than average prices, thanks to the state-wide charging network and higher public acceptance of electric vehicles.

A good selection of used cars can typically be found at franchised dealers. You will also find them advertised online by independent dealers and private owners.

## Fuel Costs

The biggest ongoing cost for a conventional car is fuel. With an electric car, the additional electricity cost is negligible.

To recharge a Nissan LEAF from completely flat to completely full takes around 25kWh (or 25 units) of electricity. Exactly how much this costs will depend on your electricity tariff, but a number of energy companies now offer specific plans for electric car owners that offer discount rates for overnight charging.

If you drive a Nissan LEAF for an average of 30 miles each day – around 11,000 miles a year – you are likely to use around 7kWh of electricity to recharge your car each night.

## Cost savings in the United Kingdom

In the United Kingdom, based on 8p per unit cost for off-peak electricity, the cost of recharging the car would be 56 pence a night, or £17 each month.

Compared to a conventional car that gives 35 mpg (7.7 km per litre, based on the imperial gallon), you are using roughly 136 litres of petrol per month to go the same distance. At a cost of £1.30 a litre, that is a total cost of £177 per month.

In this example, the cost of charging an electric car is around 9½% of the cost of putting fuel into a conventional car.

## Cost savings in the United States

In the United States, based on 11 cents a unit cost, the cost of charging the same car would be 77 cents a night; around $23 a month.

Compared to a conventional car that gives 30 mpg, you are using roughly 35 gallons of petrol per month to go the same distance. At a cost of $3.15 per US gallon, that is a total cost of $110 a month.

In this example, the cost of charging an electric car is around 21% of the cost of putting fuel into a conventional car.

# A car for free?

I've already alluded to the fact that cost of leasing a LEAF can be offset against the cost of fuel, but the figures are actually more exciting than that.

Back in 2009, I wrote an article for a national newspaper that suggested that by 2016 in Europe, and 2018 in the United States, electric cars could effectively be given to customers in return for their monthly fuel money. I argued that fuel prices were always going to rise in the long term and that electric cars were going to become cheaper. At some point there would be a crossover point where you could own a whole car for the same price as pumping fuel into a conventional one.

At that crossover point, I said, people would stop asking why they should consider buying an electric car, and instead ask why they should buy anything else. It would stand to reason that at that point there would be a big move away from conventional cars towards electric cars.

This crossover point would also increase the size of the new car market. People who could not afford a new car, because of the cost of fuelling their existing one, would be able to get a brand new car without it affecting their household budgets. They would benefit from having a fixed budget for all their car expenses without the worry of fluctuating oil prices. Older, gas guzzling cars could be taken out of use and recycled. The move away from conventional engines to electric power could potentially happen a lot faster than any of the industry pundits are predicting.

We see this rapid switch from old technology to new all the time in the computing and entertainment industry. Consider how quickly mobile phones came into our lives. Flat screen TVs. Laptop computers. MP3 players. Tablets. Wi-Fi. How the Internet changed almost every aspect of our lives. Each of these technologies started small and insignificantly. Each of them wiped out entire technologies – and industries – that came before them. You would be a brave person not to bet the same could happen with electric vehicles.

Today, I am very close to seeing my prediction come true. For me personally, it already has: after a modest initial deposit, I now pay less to lease my 2014 LEAF, and the electricity to run it, than I used to pay for fuel in my previous car. If you are doing 12,000 miles or more a year, you may find the same is true for you as well.

# Looking after your LEAF

The Nissan LEAF has proved itself to be a reliable car. As with all cars, however, you'll get the best out of it if you look after it and drive it well. Race any car like a rally car and it will wear out sooner than a well looked after car that is driven economically. The LEAF is no exception.

## Looking after the battery

The lithium-ion battery that powers the LEAF is designed to last as long as the car itself. In the United States, the standard battery warranty is eight years/100,000 miles. In Europe, the battery warranty is five years/60,000 miles. The length of this warranty demonstrates Nissans' confidence in the long life of the battery pack: they would not offer such a long warranty unless they were absolutely confident that the pack should comfortably outlast this time.

However, in the same way that an engine gradually loses performance and economy as it ages, the battery in an electric car will similarly deteriorate over time. This deterioration is so gradual that most owners will not notice it, even if they cover large distances every year.

There are ways to look after the battery and therefore extend its life. Most of these are very easy to achieve and you will probably do all these things anyway:

- In very cold temperatures, drive gently for the first five minutes of driving. Like a conventional car, electric cars don't particularly like the cold. After a few minutes of use, the batteries will begin to warm up and will perform much better.
- Use ECO mode for general driving. You still have a fair amount of power available in eco mode, but you use 10% less energy. Your range is improved and your batteries prefer it, too.
- If the battery is low, charge it up. Do not leave it unless you have no choice.

- Do not leave your LEAF with a flat battery for more than a few days. Particularly in very cold conditions.
- If ambient temperatures go below -17°C (-1°F), try to ensure the car is plugged in whenever it is not in use.
- Do not completely flatten the battery on a regular basis. It is fine to keep driving the car until the 'Low Battery warning' appears, but if you are regularly driving the car until it grinds to a halt, you will reduce the overall lifespan of the battery.
- Use the Long Life Mode (see the section on Long Life Mode below) for normal day-to-day driving if practical.
- Do not use rapid DC charging on a daily basis to recharge your LEAF. Batteries prefer to charge up over several hours where possible. Charging using a domestic power supply or a dedicated home charging point is easier for the batteries and will help extend their life.

## Long Life Mode

Long Life Mode charges the batteries to 80% and then switches off. The reason is that batteries perform at their best when they are charged between 20 – 80%. Most of the deterioration occurs outside this range. Consequently, if your day-to-day driving means you are unlikely to be worried about range, you can configure your LEAF to set the maximum charge to 80%. This can easily be overridden for the occasions when you need to go further.

On my car, I rarely need to go long distances for most of the week, but sometimes do on Thursdays and Sundays. I have set up charge timers on my car so that it switches on charging at midnight each night for the rest of the days in the week and only charges to 80%. On Wednesday and Saturday nights, I allow the car to charge up to 100% in order to give me the extra range.

It has to be stressed that using Long Life Mode is not mandatory. Most Nissan LEAF owners are blissfully unaware of its existence. If you are buying your LEAF and plan to keep it for a long time however, Long Life Mode will certainly help to protect your investment.

# The Nissan LEAF and the environment

How green is a Nissan LEAF? Are they truly environmentally friendly, or is it just greenwash? Electric car enthusiasts are always keen to point out the fact that their cars do not emit any pollution where they are being used. Detractors point to the coal-fired power station generating the electricity in the first place.

Both groups are making a valid point but, taken in isolation, both groups are inaccurate. Without looking at the whole picture, no fair assessment of the relative merits and disadvantages of different technologies and vehicle types can be made.

The US-based Automotive Science Group evaluate cars from many of the major car manufacturers to assess their environmental impact. It has a comprehensive methodology called the Automotive Performance Index (API) that takes into account the raw materials, assembly, transportation, consumers' use and end-of-life recovery. It also takes into account the social responsibilities of the manufacturer.

These assessments are carried out on electric cars, hybrid cars and conventionally powered cars alike. The API is amongst the most comprehensive and certainly the most accessible environmental assessments carried out across the motor industry today.

Based on this API, the Automotive Science Group have rated the Nissan LEAF as the top mid-sized car for Best Environmental Performance, for Best Social Performance, measuring the protection of environmental justices for all people; and finally for Best All Round Performance, with the best combined social, environmental and economic performance in class.

Of course, the biggest area of debate is about how clean our electricity supply is. It's difficult to see why electric cars are regarded as clean when there is a coal fired power station belching out industrial amounts of pollution every second. Claiming that an electric car is 'pollution free' just because it does not have an exhaust pipe is quite clearly inaccurate.

Current comparisons between conventionally powered cars and electric cars fail because they typically only measure the 'tank to wheel' emissions of the exhaust from a car. Both the European Commission and the US Environmental Protection Agency has defined a standard set of tests for measuring these emissions. They measure the emissions from the point the fuel has entered the car to the point where the energy is used. Of course, electric cars benefit significantly from this measurement because by themselves they do not pollute at all: all the pollution happens at the power station where the electricity is generated.

However, in the same way that a 'tank-to-wheel' measurement does not measure the true carbon footprint of using an electric car, neither does it measure the true carbon footprint of using a conventional combustion engine car.

For a conventional car, the carbon footprint for extracting, refining and transporting the oil needs to be taken into account, not just the carbon footprint for the emissions coming out of the tailpipe. For an electric car, the carbon footprint for getting the raw fuel, transporting it to the power station, generating the electricity and 'delivering' it to the plug needs to be taken into account.

These measurements are called 'well-to-wheel' measurements. In order to be able to make a true comparison between electric cars and combustion engine

*"Electric cars aren't green when you factor in where the electricity comes from," say the detractors...*

cars, we need to be able to identify this well-to-wheel calculation for both oil use and electricity use.

To make a proper comparison between an electric car and a conventional car, there are several measurements that need to be considered:

- Air pollution - at the tailpipe and at the electricity power station.
- How the energy is produced and transported - all the way back to where the oil is pumped out of the well and coal is extracted from the mine.
- Fuel economy.
- The environmental impact of batteries.

In a nutshell, when you take into account the sourcing and refining of oil, the full 'well-to-wheel' carbon emissions of a conventional car are around 20% higher than the official 'tank-to-wheel' emissions.

Likewise, for electricity production, the emissions for sourcing the fuel and the transmission losses add around 15% to the official carbon emissions of the power stations. You also need to factor in the carbon footprint for the construction and eventual decommissioning of the power station and for the battery use in the electric car.

Once you have taken all of this into account, a proper comparison between an electric car and a conventional car can then be made.

*..yet oil refineries are a big source of pollution, which needs to be factored into the environmental assessment for conventional cars as well.*

Based on the European measurements of vehicle fuel efficiency and of carbon dioxide emissions per km travelled (g $CO_2$/km), but based on 'well to wheel' emissions for both the electric cars and the conventional cars, here is how the Nissan LEAF stacks up against its rivals when charged up from electricity from the UK, the US, Ireland and Canada:

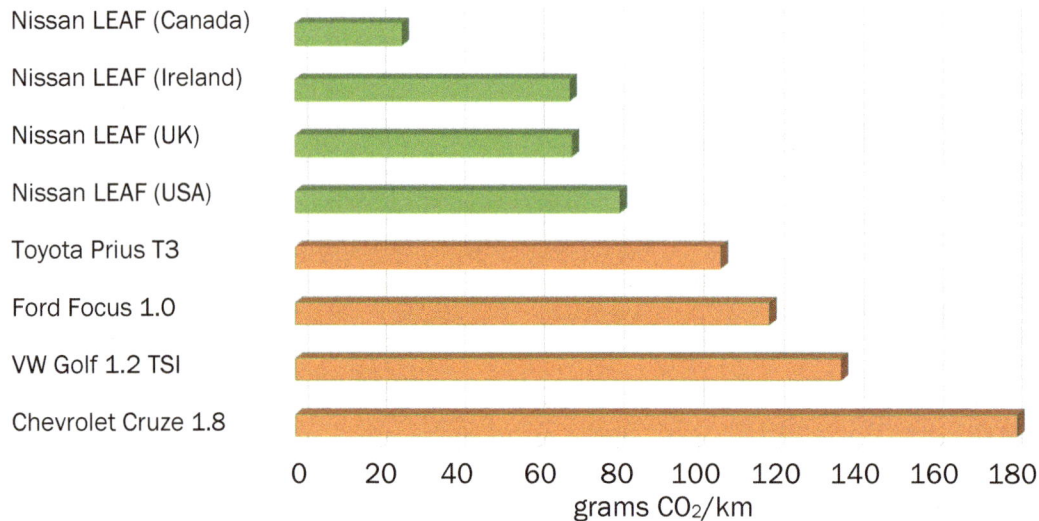

As you can see, the exact carbon footprint of the Nissan LEAF depends on the mix of generation sources for our electricity. In Canada, which has a high proportion of hydro-electric power, the average carbon footprint for driving a Nissan LEAF is around 27g $CO_2$/km. In the UK and Ireland, which has a higher proportion of coal and gas-fired power stations, the figure is around 70g $CO_2$/km, whilst in the United States, where much of the electricity comes from coal-fired power stations, the figure is around 82g $CO_2$/km.

However, all these figures are far lower than the equivalent figures for a hybrid or conventional car. For example, the Toyota Prius T3, which has official tank-to-wheel emissions of 89g $CO_2$/km, is actually responsible for 107g $CO_2$/km when the carbon impact of extracting the oil, refining it and transporting it to the service station are all taken into account. Likewise, the Ford Focus 1.0

EcoBoost, with official figures of 99g $CO_2$/km is responsible for 119g $CO_2$/km with a well-to-wheel measurement.

I go into far greater detail about the environmental impact of transportation and the benefits of electric cars in my book *The Electric Car Guide*, which is currently in its 2015 edition. This book goes into far more detail about where the figures that I use in this book come from, more about how our electricity is generated and how oil is refined into fuel.

So, based on theoretical data, the Nissan LEAF can be shown to be far more environmentally friendly than conventional alternatives. But how about in practice? In the next chapter, I explain exactly how to prove the environmental benefits of the LEAF in practice.

# Real world economy and the environmental test

In the real world, with real-world driving, it seems virtually impossible to get the fuel economy figures and range that the manufacturers claim for their models, particularly if you take the New European Driving Cycle (NEDC) measurements as a benchmark. Whether you are talking about combustion engine cars or electric cars, the figures appear to have little relevance in the real world other than to work as a comparison with other makes and models.

In an electric car, however, you can create your own test measurements extremely easily. Simply record your distance travelled and then measure the amount of electricity used to recharge the car using a plug-in watt-meter.

## The Test

In order to measure real world economy for the Nissan LEAF, I carried out a driving test along a fixed route using a 2014 Nissan LEAF.

In order to provide a useful comparison, I then tested two similarly sized cars, a Toyota Prius and a Ford Focus 1.0-litre eco-boost, driving the same route, in order to identify real-world fuel economy and comparative carbon footprint figures for each type of vehicle.

I have carried out this test before using other vehicles and have established a regular test route around Coventry in the United Kingdom. The total distance is 14.1 miles, comprising of 2½ miles of dual carriageways and 4½ miles of busy inner-city roads. Ambient temperature was between 12–14°C (between 54 and 57°F)

## The Carbon Calculations

In order to measure the carbon calculations as accurately as possible, I need to measure the carbon footprint of the electricity I use to charge up the Nissan LEAF. I also need to take into account the sourcing of the fuel, the carbon impact of the power station itself, and the transmission losses of the electricity between the power station and the plug.

Likewise, to make an accurate comparison, I need to measure the carbon footprint for the sourcing, refining and transporting of the fuel, none of which is measured in the official carbon measurements for internal combustion engine cars.

## Test Validity

It is worth stressing that these tests have not been independently verified by any scientific establishment. Consequently, these tests can only ever be used as an indication of relative fuel economy and carbon emissions.

I also feel that it is important that my test could be repeated by anyone else using their own cars and their own routes, and the tests have been simplified in order to achieve this.

All the information and calculations I used in order to carry out my tests are included within this chapter, whilst more detailed information on how electricity is generated and oil is refined is available in my other book, *The Electric Car Guide* which is currently in its 2015 Edition. If a university or a scientific establishment wishes to carry out similar tests in a controlled environment and would like to discuss my test methods, I can be contacted through the *Ask Me a Question* page of the website *www.TheElectricCarGuide.net*.

## The Nissan LEAF

The Nissan LEAF I used was my current car, a 2014 LEAF Acenta, which has travelled 4,300 miles. The car was plugged in with a charge timer set to switch on the charging at midnight. The amount of electricity used was measured using a watt-meter and this was then multiplied by the average carbon footprint for the UK for the period when the cars were on charge, using figures supplied by the National Grid.

The carbon footprint calculation for the electricity took into account the carbon impact for sourcing and transporting the fuel used at the power stations, the production of the power itself and the average transmission losses of the power as it is delivered from the power station to the car. It also factors in the carbon footprint for the construction and eventual decommissioning of the power stations themselves (a particular issue with nuclear power which is rated as very low carbon until the decommissioning impact is taken into account).

These carbon calculations and real time information can be seen on the Electric Car Guide website (www.TheElectricCarGuide.net). A free smartphone app called ECO2 has also been developed to complement this book, which provides this information in real time on your mobile phone. You can download this app for free by visiting Google Play or the iStore.

## The Toyota Prius and Ford Focus

The two comparison cars that I chose were the Toyota Prius T3 and the Ford Focus 1.0 litre EcoBoost.

The Toyota Prius T3 was the original hybrid and still provides good economy figures today. Similar in size and cost to a Nissan LEAF, the Prius gives excellent fuel economy and can drive in electric only mode for around 1½ miles. 0-60 takes around 11.8 seconds.

The Ford Focus 1.0 litre EcoBoost represents the very latest conventional engine technology, with an ultra-lightweight, 3 cylinder turbocharged one litre engine, producing 99bhp. 0-60 takes around 12.5 seconds.

Both of these cars are economical family cars that produce low levels of carbon emissions. The manufacturers' own $CO_2$ footprint figures show that, in official European tests, the Toyota Prius T3 produces a tank-to-wheel footprint of 89g

$CO_2$/km, while the Ford Focus produces a tank-to-wheel footprint of 99g $CO_2$/km.

These figures only reflect the tank-to-wheel emissions, not the well-to-wheel emissions. For our tests to be comparative to the electric car tests, well-to-wheel calculations have to be used.

In order to calculate the carbon footprint in real conditions, I filled the fuel tank at the start of the test. I then measured the fuel economy in litres at the end of each test by refilling the fuel tank. I calculated the $CO_2$ footprint based on the amount of fuel used, using the 'well-to-wheel' $CO_2$ fuel figures. More information about these well-to-wheel figures is published in The Electric Car Guide – 2015 Edition.

## Test results from the Nissan LEAF

| | 2014 Nissan LEAF |
|---|---|
| Distance Travelled: | 14.1 miles (22.6 km) |
| Total Electricity Used: | 3.37kWh |
| Estimated Cost: | 28p (around 42¢) |
| Average $CO_2$ during charging: | 405g $CO_2$/kWh |
| $CO_2$ electricity usage: | 60g/km |
| Estimated $CO_2$ battery usage: | 2g/km |
| Total $CO_2$/km: | 62g/km |
| Total $CO_2$ for journey: | 1.40kg |

## Test results from the Toyota Prius and Ford Focus

| | Toyota Prius T3 | Ford Focus 1.0 EcoBoost |
|---|---|---|
| Distance Travelled: | 14.1 miles (22.6 km) | |
| Total Fuel Used: | 1.01 litres | 1.07 litres |
| Total Fuel Cost: | £1.10 | £1.18 |
| $CO_2$ per litre of fuel | 2789g $CO_2$ per litre | |
| Official $CO_2$ tank-to-wheel: | 89g $CO_2$/km | 99g $CO_2$/km |
| Actual $CO_2$ tank-to-wheel: | 125g/km | 132g/km |
| $CO_2$ well-to-wheel: | 151g/km | 159g/km |
| Total $CO_2$ for journey: | 3.41kg | 3.59kg |

As you can see, the carbon footprint figures that I achieved in my test were significantly higher than the official figures. The official tests have been carried out in laboratory conditions, typically performed on a roller test bench at an ambient temperature of between 20 – 30°C (68 – 86°F). Tests are designed to emulate driving on a completely flat road with an absence of wind. Inevitably, there is significant difference between tests carried out in these conditions and real-world driving.

## Side-by-side analysis: well to wheel measurements

| | Nissan LEAF | Toyota Prius T3 | Ford Focus |
|---|---|---|---|
| Fuel cost | £0.28 | £1.10 | £1.18 |
| $CO_2$ per km | 62g | 151g | 159g |
| $CO_2$ for journey | 1.40kg | 3.41kg | 3.59kg |

# What if my Nissan LEAF was powered by coal?

When a Nissan LEAF is powered by a coal-fired power stations, the carbon footprint is significantly higher than when they are powered by any other source.

Using the figures of 977g/kWh for coal-fired power, which includes the mining and transportation of the coal and the energy losses between the power station and the home, this is what the $CO_2$/km would look like if I charged up using coal power:

|  | Nissan LEAF | Toyota Prius T3 | Ford Focus |
|---|---|---|---|
| Fuel cost | £0.28 | £1.10 | £1.18 |
| $CO_2$ per km | 145g | 151g | 159g |
| $CO_2$ for journey | 3.29kg | 3.41kg | 3.59kg |

Based on these figures, the worst case scenario shows that a Nissan LEAF is between 4 – 10% more carbon efficient than the most efficient internal combustion engine cars.

Of course, these figures have not been independently verified and can only ever be used as an indication of relative fuel economy and carbon emissions. However, as a real world demonstration of the cost savings and carbon savings, they are easily repeated. Why not have a go yourself?

# A final word

Electric cars are a radically different and exciting new technology that have now matured enough to enter the mainstream. Today, they are a practical alternative to conventional cars and have significant environmental and economic benefits. For many people, they are the ideal vehicle, providing quiet, smooth and practical motoring, with the convenience of fuelling from home at the fraction of the cost of pumping fuel at a service station.

If I've encouraged you to go out and try a Nissan LEAF for yourself, that's great. I've achieved what I set out to do. Likewise, if you have read the book and come to the conclusion that a LEAF is not for you, this book has also served its purpose. Far better to spend a small amount of money on a book than spend a lot of money buying the wrong car. If you already own a LEAF, hopefully this book has given you a little more insight into your car and how to get the very best out of it.

Finally, if you've enjoyed this book, or even if you haven't, feel free to get in touch. If you have any questions about the Nissan LEAF, or in electric cars in general, or if you have suggestions on how I can improve this book, I would be delighted to hear from you. I can be contacted through the *Frequently Asked Questions* page of the Electric Car Guide web site. Find me at www.TheElectricCarGuide.net.

All the best,

Michael Boxwell
April, 2015

# Also by Michael Boxwell

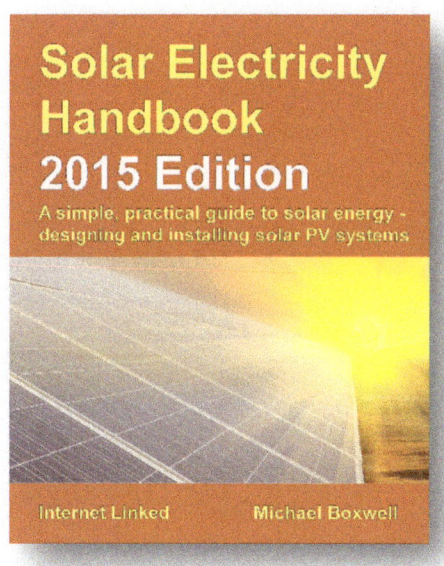

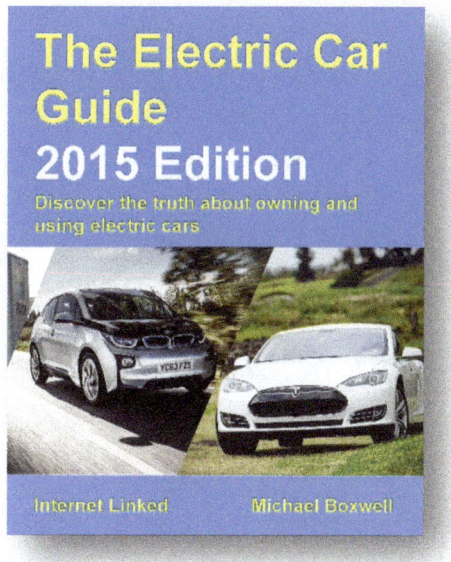

The Solar Electricity Handbook is the world's best-selling guide to solar power. Assuming no previous knowledge, the book explains how photovoltaic panels work and how they can be used. It provides a step-by-step guide to successfully design and install a solar energy system from scratch.

Michael has been involved with solar energy systems since 1999. His designs have been used around the world and he has advised both the UK and US governments on solar energy policy.

The Solar Electricity Handbook is published by Greenstream Publishing and available from Amazon and all good book sellers.

The Electric Car Guide explains exactly what it is like to own and use an electric car on a day-by-day basis. Author Michael Boxwell has been driving electric cars since 2006 and previously ran one of the largest electric car clubs in the world. He has worked with various car makers on the design and development of electric car technology since 2009.

By the time you have finished reading The Electric Car Guide, you will understand what it is like to own, use and live with an electric car. You'll understand the arguments both for and against electric car use and you will know whether an electric car is suitable for you.

Lightning Source UK Ltd.
Milton Keynes UK
UKHW050842301218
334718UK00003B/51/P